OTHER TITLES IN THE SERIES

Alternative Kilns
Ian Gregory

Ceramics with Mixed Media
Joy Bosworth

Ceramics and Print
Paul Scott

Coiling
Michael Hardy

Colouring Clay
Jo Connell

Crystalline Glazes
Diane Creber

The Electric Kiln
Harry Fraser

Glazes Cone 6
Michael Bailey

Handbuilding
Michael Hardy

Impressed and Incised Ceramics
Coll Minogue

Kiln Building
Ian Gregory

Large-scale Ceramics
Jim Robison

Lettering on Ceramics
Mary White

Oriental Glazes
Michael Bailey

Paper Clay
Rosette Gault

Porcelain
Jack Doherty

Raku
John Mathieson

Resist and Masking Techniques
Peter Beard

Setting Up a Pottery Workshop
Alistair Young

Single Firing
Fran Tristram

Slipcasting
Sasha Wardell

Stoneware
Richard Dewar

Soda Glazing
Ruthanne Tudball

Throwing Pots
Phil Rogers

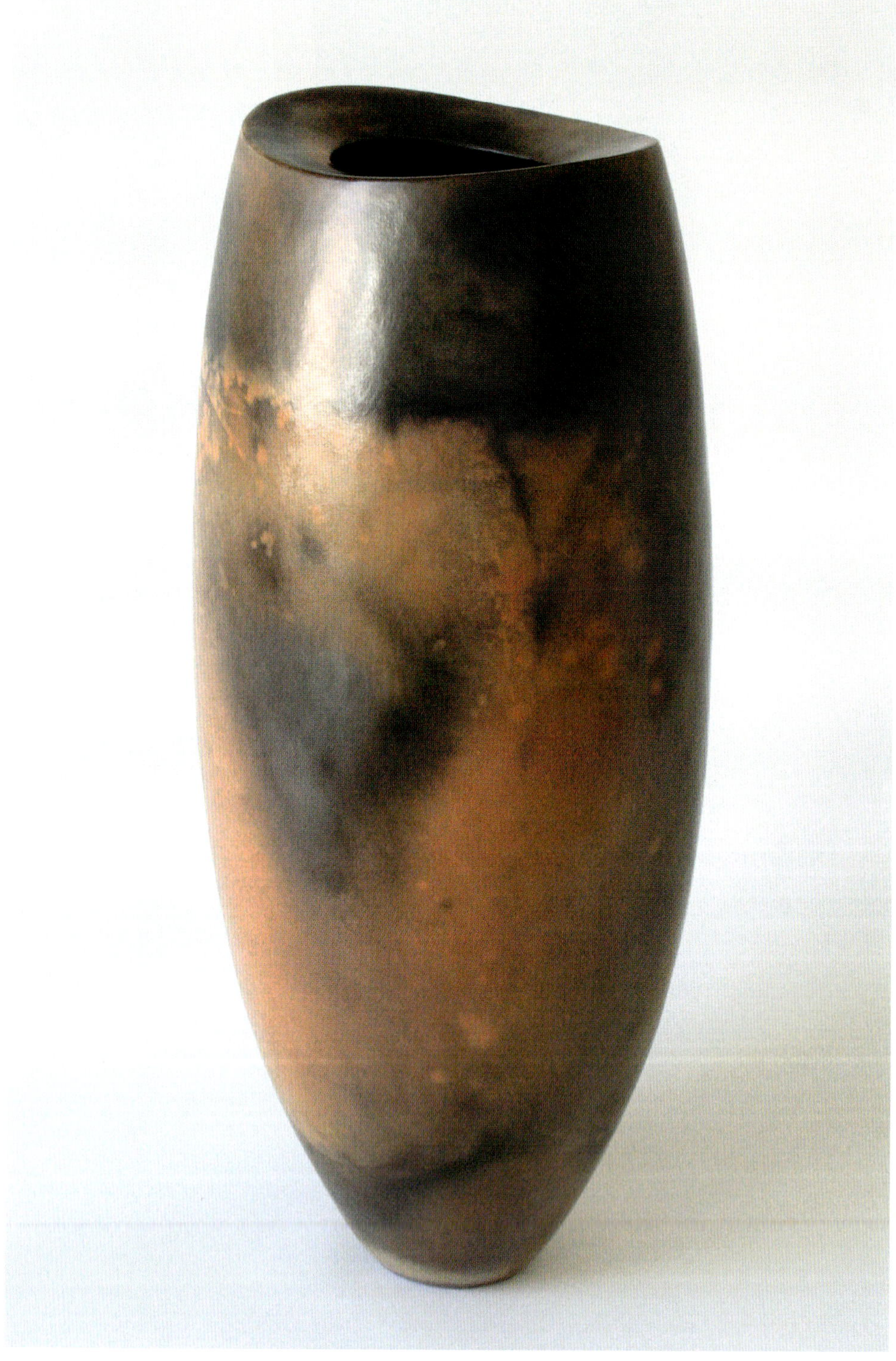

LOW FIRING
&
BURNISHING

Sumi von Dassow

HERBERT PRESS
LONDON • OXFORD • NEW YORK • NEW DELHI • SYDNEY

The American Ceramic Society · Ohio

COVER (FRONT) *Horsehair Urn* by Sumi von Dassow, 2006. Ht: 35 cm (14 in.), dia: 30 cm (12 in.).
COVER (BACK) *Large Bowl* by David Greenbaum. Dia: 59.5 cm (23.5 in.). *Photo by Randy Batista.*

FRONTISPIECE *Amphora* by Gabriele Koch, 2007. Ht: 54 cm (21 in.). Modified T-material, coiled, burnished slip, smoke-fired. *Photo by Kelpie.*

HERBERT PRESS
Bloomsbury Publishing Plc
50 Bedford Square, London, WC1B 3DP, UK
29 Earlsfort Terrace, Dublin 2, Ireland

BLOOMSBURY, HERBERT PRESS and the Herbert Press logo are trademarks of Bloomsbury Publishing Plc

First published in Great Britain 2009

Published simultaneously in the USA in 2009 by
The American Ceramic Society
600 N. Cleveland Ave., Suite 210,
Westerville, Ohio 43082

A catalogue record for this book is available from the British Library

Library of Congress Cataloguing-in-Publication data has been applied for

ISBN: PB: 978-1-912217-58-8;
US ed: 978-1-574982-93-0

4 6 8 10 9 7 5

Series: Ceramics Handbooks
Typeset in 10 on 12.5/2pt Photina
Book design by Susan McIntyre
Cover design by Sutchinda Thompson

Printed and bound in India by Replika Press Pvt. Ltd.

To find out more about our authors and books visit www.bloomsbury.com and sign up for our newsletters.

Contents

Acknowledgements 6

Introduction 7

1 Burnished pottery in history 11

2 About burnishing 17

3 Burnishing with terra sigillata 35

4 Smoke-firing and black-firing 51

5 Pit-firing 63

6 Saggar firing 74

7 Raku firing techniques: horsehair and naked raku 89

8 Finishing touches 101

Bibliography 106

List of suppliers 107

Contributing artists 108

Glossary 109

Index 111

Acknowledgements

How is it possible to acknowledge everyone who has helped shape my journey in clay? I will begin with my high school pottery teacher, Ed Pincus, who gave me, a straight-A student, the first C of my life in my first pottery class, for my attitude. He wouldn't let his students work on the wheel for the entire first semester, and I was itching to throw pots. I stuck with it just to show him I didn't deserve that grade! Then Howard Kottler and Patti Warashina at the University of Washington, who pushed me to branch away from wheel-thrown pots.

Then Janet Lohr who taught a class called 'Pottery of the Americas' at the Fort Mason Center in San Francisco. I took it because it was the only class at a time that was convenient for me, figuring I'd ignore her instruction and work on the wheel instead – the bull-headed young person I was then was still obsessed with throwing – but I was too polite to do that so I learned how to coil-build and burnish, and fell in love with burnished pottery. I am still using the clay she introduced me to, Navajo Wheel.

Then on to San Francisco State University, where David Kuraoka introduced me to pit-firing on the San Francisco beach, and allowed me to teach the other students how to burnish. I didn't really intend to become a potter or ceramic artist, but I registered as an art major so I could have a locker in the art building until I figured out what I really wanted to major in. I never did figure out anything else I wanted to do, so here I am, still doing pottery.

I must acknowledge the impact of my parents and grandparents, all writers and teachers, in shaping my desire to both teach and write. Bill Jones, editor of *Pottery Making Illustrated* magazine published many articles of mine, allowing me to get the hang of writing about pottery-making. Alison Stace at A & C Black has put up with all my blunders in the process of writing this book. And finally, of course, without a husband who encouraged and supported my desire to develop my pottery, I would have had to get a real job instead.

Introduction

Potters who burnish are often asked, 'what glaze is that?' by curious admirers of their work. Non-potters naturally assume that all pottery is glazed, and the glossy surface of a burnished pot seems like a different and intriguing sort of glaze. Though glazed pottery can be brighter and more colourful, a burnished pot has a glow from within and a warmth that glazed pottery doesn't possess. The difference which non-potters sense without knowing it – and which fascinates potters – is that the surface of a burnished pot doesn't wear a coat hiding the clay itself from view. Glaze is glossy and reflective, but the reflecting surface consists of a millimeter or so of glass covering the clay. Underneath this layer of glaze the rough stony clay is always perceptible, even if not always visible. A burnished pot can have a surface just as glossy and reflective as any glaze, but behind this glorious surface there is no hidden roughness. Even the feel of a burnished pot is seductive: while a glazed pot feels hard and cold, a burnished pot seems warm and almost soft to touch. Potters who burnish get used to seeing their customers handle the pots, turning them in their hands and stroking the surface. This is a common and unconscious response to the sensuousness of burnished pottery.

While nowadays most pottery, certainly most functional pottery, is glazed to make it sanitary and non-porous, throughout history potters all over the world have taken advantage of the beautiful possibilities inherent in clay itself when not covered with glaze. When the use of clay evolved into the making of pots, the first pottery was unglazed. While in some places potters gradually developed the technique of coating pots with glaze, in other parts of the world the traditional unglazed surface is still used today. Burnishing an unglazed surface is for these traditional potters a technological advancement over leaving it rough, as a burnished surface is somewhat more watertight than an unburnished one. Today, of course, a potter or ceramic artist is free to choose from a wide range of surface finishing techniques to find the one most suited to his or her artistic vision. Those who seek bright colour, durability or vitrified ware, will choose to use glaze. Others, who want a hard, stony surface, may leave the clay bare of glaze and rough-textured. Those whose vision demands a more organic look and a softer surface, may choose to burnish.

Burnished pottery doesn't need to be fired in any particular way, except that a burnished surface cannot endure the high temperatures necessary to melt glaze without changing, and becoming dull. In cultures where burnished pottery is still made in traditional ways, the work is often fired in a bonfire, leaving marks from the burning material on the surface of the pottery. Such random smoke-markings so enhance and enrich much burnished pottery that many modern ceramic artists

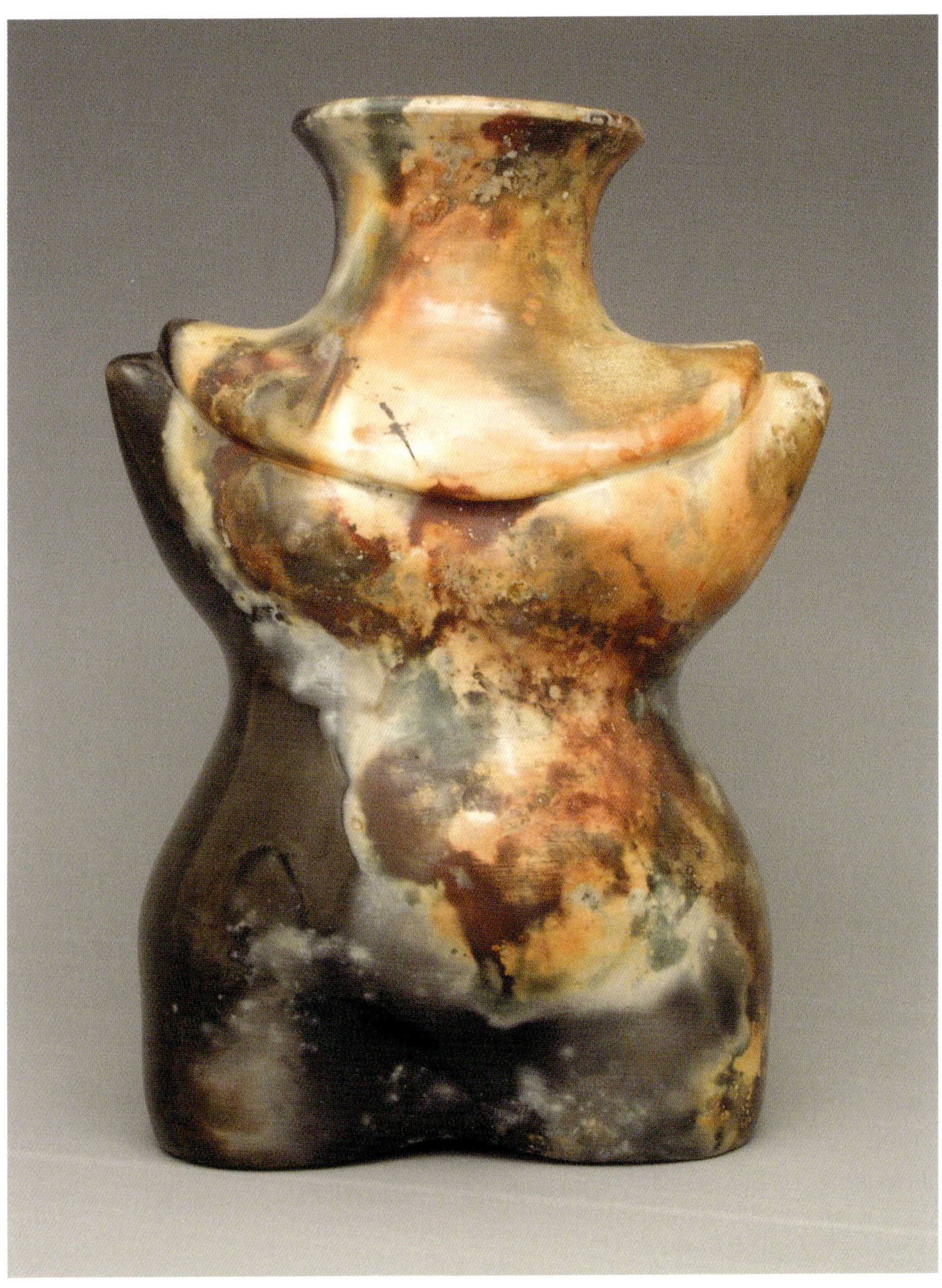

Pit-fired Torso by Sumi von Dassow, 2006. 30 x 20 x 15 cm (12 x 8 x 6 in.). Wheel-thrown and altered from B-Mix clay (white stoneware); terra sigillata sprayed on and burnished with a chamois-leather; bisque-fired to cone 010; pit-fired with salt, copper carbonate and salt-soaked materials. A guitar string wrapped bandolier-style around the torso left traces such as the diagonal line on the collar. *Photo by Sumi von Dassow.*

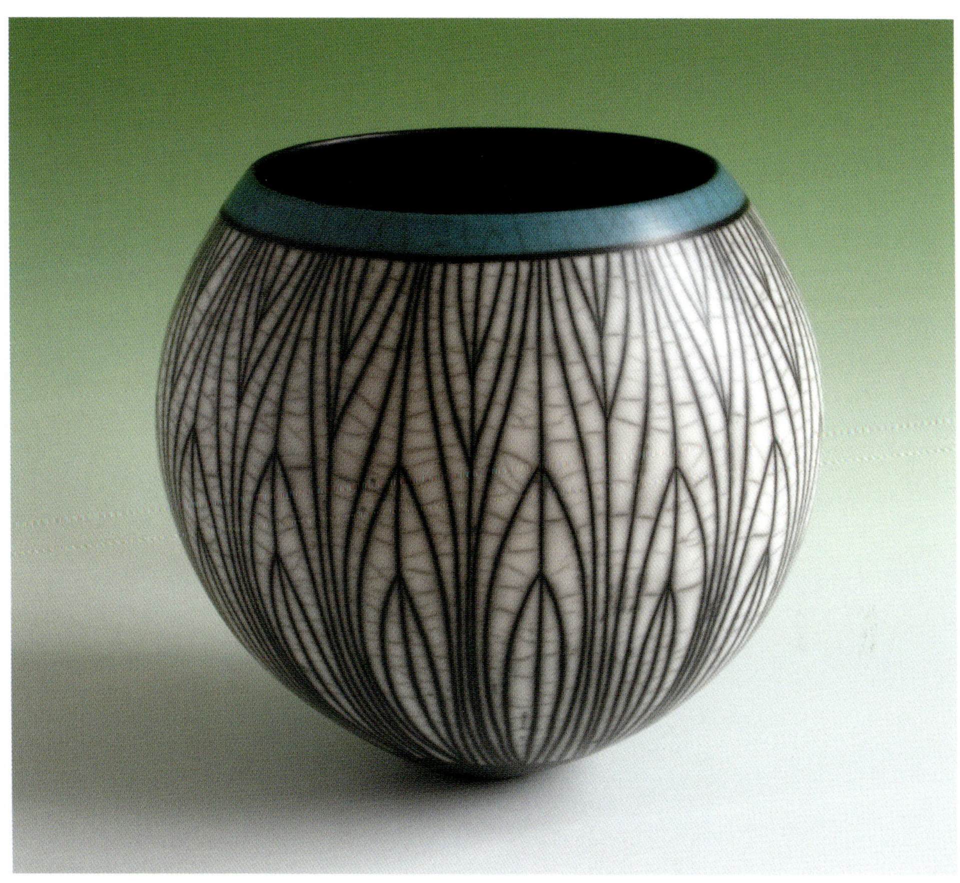

Bowl, by Ashraf Hanna. Ht: 26 cm (10¼ in.). Handbuilt, painted with white slip and slip coloured with turquoise stain, then burnished and fired using the naked raku process.

who burnish use some form of smoke-firing to finish their work. At the same time, since pit-firings and sawdust firings must generally be done with unglazed pots, many potters find burnishing to be the best way to give a finished look to pots fired with these techniques. Thus the marriage between burnishing and smoke-firing is a match created from mutual attraction, not necessity.

The aim of this book is to examine the partners in that marriage. Many images of burnished pottery have been included to allow you to savour and appreciate the versatility of this technique. Various methods of burnishing are presented so that you may choose the most appropriate technique for your particular application. And several approaches to pit-firing, saggar-firing and other types of smoke-firing are described and illustrated to offer a springboard for creativity. My hope is that potters and non-potters alike will come to a greater understanding and appreciation of both burnishing and smoke-firing by reading this book.

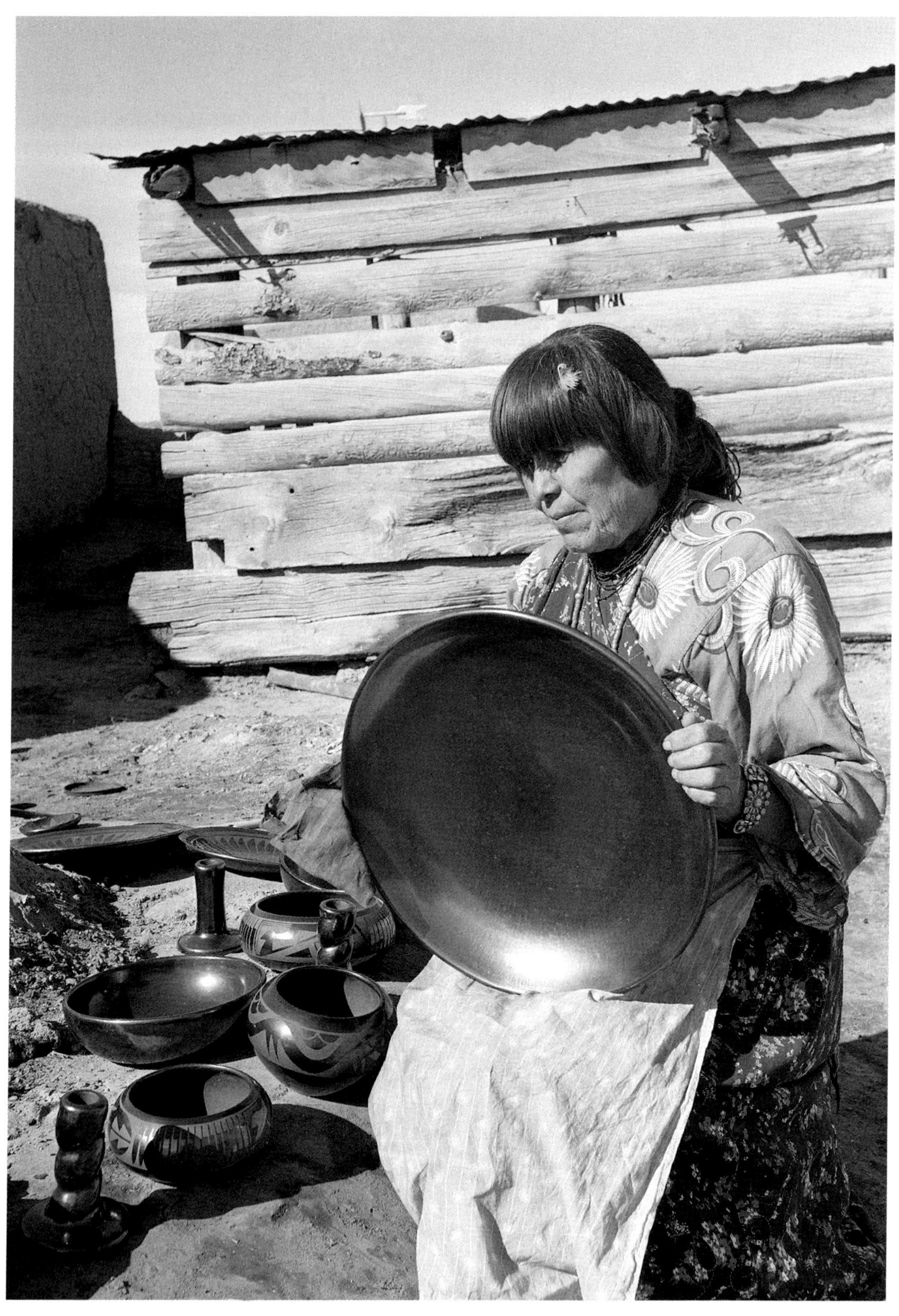

Maria Martinez holding finished pottery, San Ildefonso Pueblo, New Mexico, c.1950.
Photo courtesy of Palace of the Governors, Santa Fe.

Chapter 1

Burnished pottery in history

The development of pottery parallels the development of civilization. Nomadic hunter-gatherer societies tend not to make pottery, for pottery-making requires enough time in one place to dig and prepare clay, form pots and fire them – and pottery is heavy to carry. When communities begin to develop a sedentary, or partially sedentary lifestyle, then pottery-making becomes practical; and when a community develops agriculture and begins to herd animals, pottery becomes necessary to store the produce. In an agrarian society potters typically work the land most of the year, along with everyone else, and produce pottery only in the fallow season. When potters work part-time and make a relatively small number of pots, handbuilding and open-pit firing are ideal techniques because the technology is simple and their mastery doesn't require many years of practice.

In general, it is only when a civilization reaches a critical population mass that the production of food, clothing, domestic ware and other tasks become specialized in the interests of efficiency; and it is only when these tasks become specialized that potters turn to the use of the wheel and the kiln. These tools facilitate the relatively quick and efficient production of durable ware, but they require long periods of training and specialized knowledge compared to coil-building, burnishing and open-pit firing. Thus, as a general rule, handbuilding and low-temperature, low-technology firing persist until the population density of a region necessitates the development or adoption of the potter's wheel and kiln.

All over the world wherever clay is found its usefulness was discovered when human culture reached the appropriate stage of development – obviously predating the discovery of glaze. Unglazed pottery was the norm throughout most of history, and in many parts of the world traditional potters continue to make unglazed pottery for domestic use or for sale. Because glazes require relatively high temperatures, they must be fired in a kiln, so the pottery of agrarian and lightly populated societies tends to be unglazed even today. The earliest pottery made in any region is always handbuilt and any decoration is rendered in the earth tones of clay slip and mineral oxides.

Burnishing – polishing a pot by rubbing the surface with a smooth object – is one of those techniques that any student of pottery discovers accidentally, and the advantages are obvious. A pot which has been burnished is less porous, and thus more suitable for food and water storage because the burnishing process compresses the surface and packs the clay particles closer together. A burnished surface is also attractive, sensuous to the touch, and provides a smooth ground for painted decoration. Thus it is hardly surprising that burnished pottery shows up all over the world, and that the archaeological remains of many civilizations bear a familial resemblance – ancient pottery from China or the Mediterranean region almost seems more closely related in form and decoration to native African or American pottery, than to modern pottery from those regions. But now that modern pottery has come full circle to rediscover the beauty of burnished pottery, the history of unglazed pottery around the world is of interest to the modern ceramic artist.

China

China's legacy of skilfully crafted and decorated pottery dates back to at least the 5th millennium BC. Neolithic Chinese pottery is handbuilt, though by the 3rd millennium BC the wheel came into use. Even at such an early date the ware was apparently fired in something more closely resembling a kiln than an open pit or bonfire, and it was often burnished and painted in slip with geometric designs. Other early burnished Chinese pottery was reduction-fired to blacken it, a technique used to great effect in the production of the famous Pueblo Indian black pottery from the American Southwest. In fact, it appears that in refining their kiln design to improve the quality of this black ware, Chinese potters embarked upon the path that led to the eventual development of reduction-fired glazed stoneware. Glazes began to appear some time in the 2nd millennium BC, and from that point on Chinese potters never abandoned the pursuit for higher temperatures and more refined glazes. Sadly for admirers of burnished pottery, Chinese potters ceased to make unglazed work after AD500 or so, though in other parts of Asia handbuilt and unglazed pottery is still made and used.

The Mediterranean

Ancient Greece was producing some of the world's most famous unglazed pottery just at the time when Chinese potters were beginning to turn away from unglazed ware. The process of making Greek red and black pottery involved a sophisticated use of very simple materials, and a carefully controlled firing schedule. Pottery production, including the use of the wheel, was well developed in the Mediterranean region by about 3000BC. Handbuilt and wheel-thrown pottery, elaborately decorated with geometric patterns or stylized representations of animals or plants in painted slip, was produced throughout the region, including North Africa, the islands of the Mediterranean and the Greek mainland. For several hundred years, beginning around 1000BC, slip-decorated Greek pottery was a highly valued art form, traded widely throughout the area, and advancing technically through that period of time until it became highly refined in technique and artistic merit. Though the glossy black of this pottery is often referred to as a glaze, it was actually a fine slip made from the clay body used to form the pot, applied over a burnished clay surface. Beginning with the technique used also by the ancient Chinese of firing in reduction to produce black ware, the Greek potters discovered that, by allowing air into their kilns as they cooled, they could cause unslipped portions of their ware to reoxidise to the clay's original red colour.

Unfortunately, as the population of the region grew, so did the demand for more easily produced, less highly decorated pottery, and the heyday of Greek painted pottery came to an end by 200BC. Because the firing was simpler, potters began to make red-slipped ware and abandoned black painted decoration and reduction firing. Today, we refer to the type of slip used by potters of this region as terra sigillata, though originally this term referred to Roman pottery decorated with stamped figures and coated with fine slip. The growing population of the Mediterranean world demanded mass production of pottery for everyday use, and terra sigillata ware was easier to produce in great numbers than the painstakingly decorated pottery which preceded it. The production of this ware persisted until as late as around AD1000 in North Africa, but then glazed ware – even less porous and easier to clean than terra sigillata – won the battle throughout Europe and the Mediterranean region.

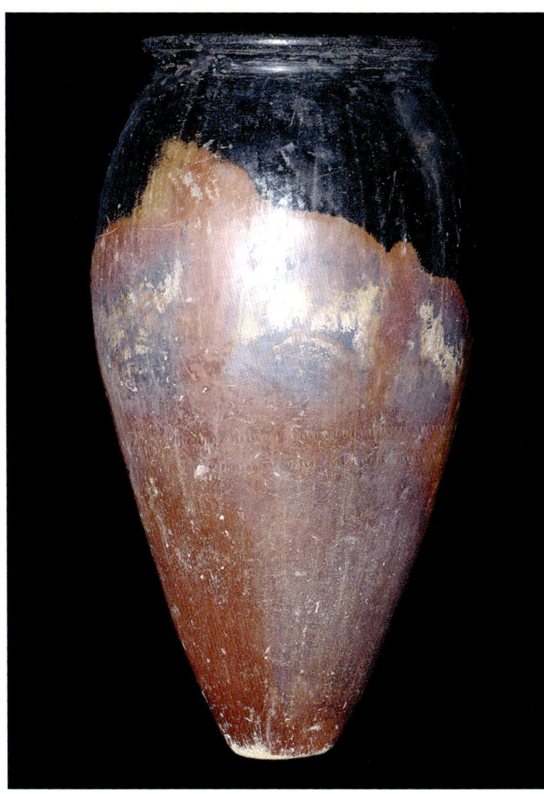

Burnished Black-topped Redware Jar. Ht: 31.5 cm (12½ in.), w: 17.7 cm (7 in.) (max.). From a tomb at Abydos, Egypt, Late Predynastic period, Naqada II, around 3200BC. Coil-built, coated with slip, and burnished with a smooth object such as a pebble. The black top is achieved by firing it upside-down in combustible material. *Photo © The Trustees of the British Museum.*

Africa

Other than in the cities of North Africa, where the culture was and is intimately connected to that of Mediterranean Europe, traditional pottery in Africa remains unglazed and fired in open bonfires to this day. This is largely because until the modern era the population of southern Africa never grew so crowded as to demand the mass production of pottery. Thus, in many parts of Africa pottery is still produced by individual villagers who work on a small scale. In some places, each woman makes her own pots, but usually the potter creates the vessels needed by the members of her tribe or village during the dry season, tending her fields the rest of the year. Though African pottery is made with simple techniques, the tradition is as old as any in the world, dating back as far as the 8th millennium BC.

The fact that glazes aren't used doesn't make African pottery unsophisticated – far from it, for much African pottery is elaborately decorated with impressed or carved texture or intricately painted line designs, any of which might be obscured by glaze. In fact, the surface decoration of African pottery is much more than an effort to make things prettier – often the purpose of the vessel, or the position in society of the user, is reflected in the decoration of the pot. In European cultures there is a linguistic connection between pottery and the human form – for instance, a person may be referred to poetically as a 'vessel' or the body as 'mortal clay'. Another manifestation of this connection is our habit of referring to parts of a pot as foot, belly, shoulder and so on. In Africa, this connection between pottery and the human body is a more profound element of the culture, and African pots often feature faces, or take the form of a figure, or incorporate elements of skin scarification patterns.

The free use of modelling and surface texture developed precisely because African potters were never constrained by the use of a wheel to form circular pots. African potters have thus had a head start on European potters in their efforts to break away from utilitarian pottery traditions and turn pottery into an art form. Straddling two worlds in this post-colonial era, European-educated African potters, exploring the techniques of their native heritage, create beautiful and thought-provoking work that is appreciated by collectors everywhere.

North and South America

The history of pottery development in the Americas is similar to that of African pottery, though it seems to be a much younger art, only appearing about 2000 years ago in North America and perhaps 1000 years earlier in South America. In pre-colonial America, as in southern Africa, the population tended to be less concentrated and more agrarian than in Asia and Europe, and evidence suggests pottery in the Americas has always been produced, as it still is today, by individual part-time potters, mostly women. As in Africa, Native American potters still dig and prepare their own clay in addition to creating and firing their pots, in between taking part in other activities of the community. All pre-colonial pottery in the Americas was hand-built, much of it was burnished and firing was done in open pits or in shallow pit-kilns. These methods of production persist to this day. As with African pottery, the use of handbuilding techniques blurs the line between pottery and sculpture, and elaborate decoration using slips and oxides is common throughout the Americas. Though potters in the Americas often produced pots in the form of animals or humans, as in Africa, in general American pottery features less textural decoration than does African pottery. Instead, Native American potters have refined the art of burnishing and decorating with coloured slips to a greater degree than anywhere else in the world.

In many parts of Central and South America Indian potters still rely on traditional methods to produce the vessels they need for cooking and food storage. Indeed, unglazed water vessels are quite practical in hot climates where the slight porosity promotes an evaporative cooling effect that keeps the water inside at a pleasing temperature. In North America, Pueblo Indian potters have developed their traditional methods of production to the point where their world-famous vessels fetch thousands of dollars, rival the work of any modern ceramic artist in skill and artistry, and are featured in major museums everywhere in the world.

In the modern era, when handmade pottery is an art-form rather than a necessity for most of the peoples of the world, handmade pottery traditions form a valuable reservoir of techniques and ideas from which artists all over the world draw inspiration. The line between traditional pottery and ceramic art has become blurred: as traditional pottery-making techniques grow impractical for more and more of the world's population, they become all the more attractive to the artist precisely because they are an antidote to sterile mass-production methods. The sculptural forms of pre-Columbian Incan and Aztec pottery, the stylized animals and geometric patterns decorating Mimbres or Anasazi pottery, and the voluptuously burnished surfaces of Pueblo pottery, would all fit in better with any exhibition of contemporary ceramic art than would most people's wedding china.

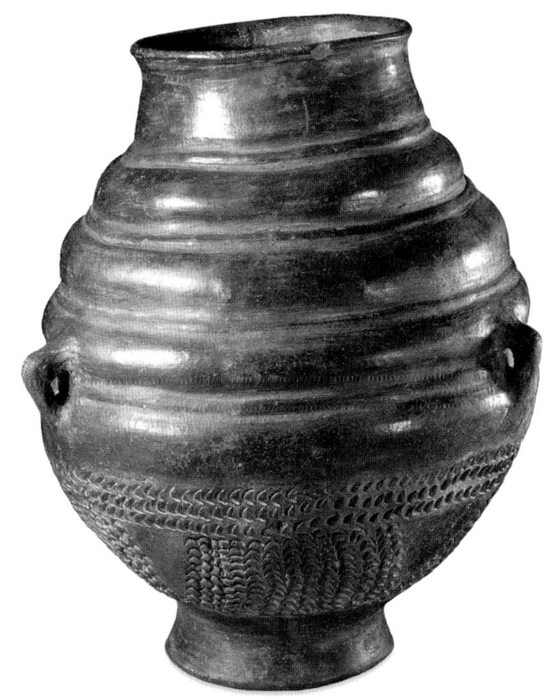

Burnished Black Pottery Jar. Ht: 20.5 cm (8 in.). Germanic, later 3rd century AD, Coil-built, burnished, and fired in a bonfire smothered with turf to blacken it. Though the techniques used to make this pot are native to the region, the pot shows the evident influence of Roman culture, imitating the form and sheen of a metal jar. *Photo © The Trustees of the British Museum.*

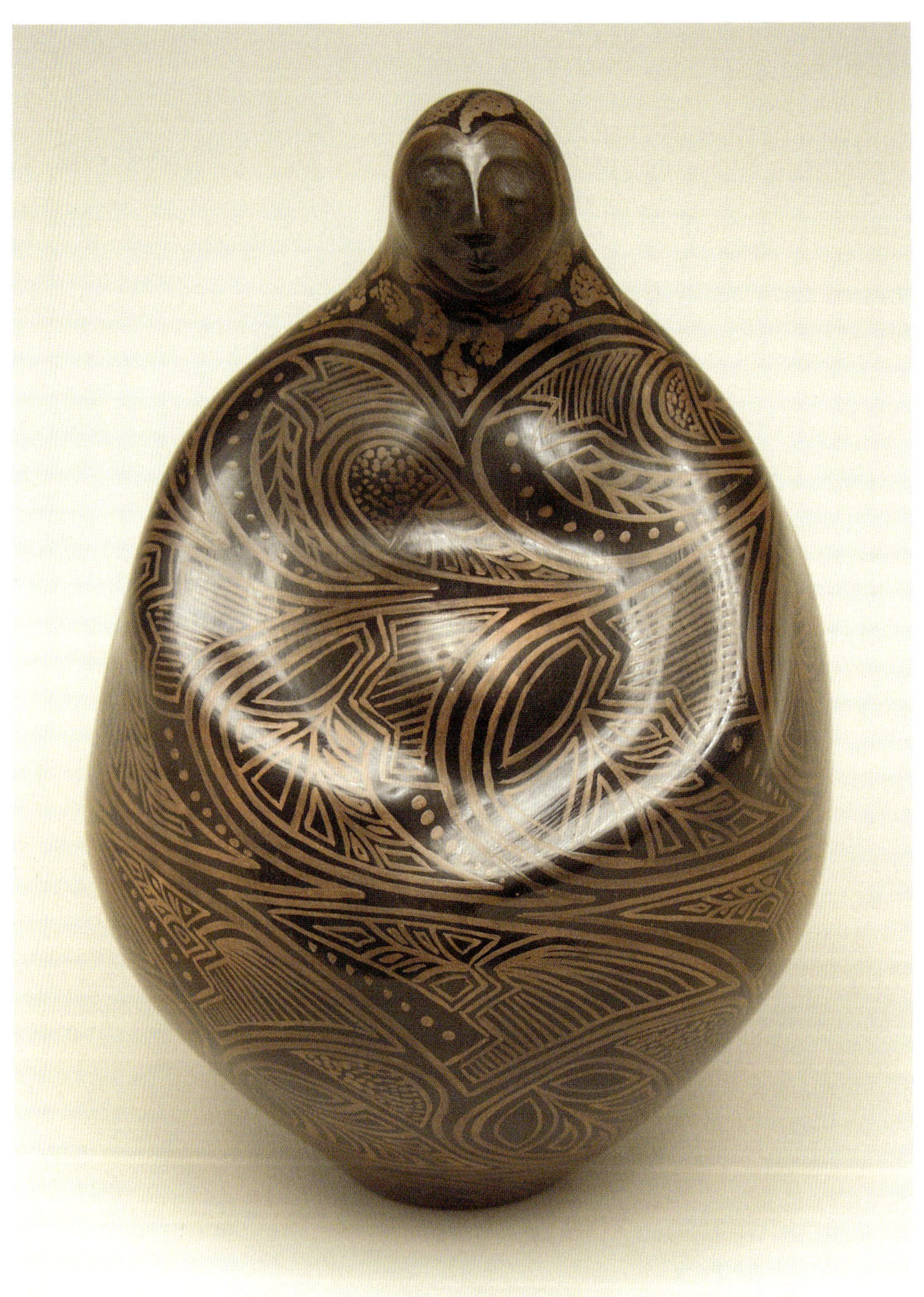

Not Just An Empty Vessel by Sumi von Dassow, 2004. 33 × 23 × 20 cm (13 × 9 × 8 in.). Coil-built from red stoneware (Navajo Wheel Clay from Industrial Minerals Co.), burnished with a stone, decoration painted on using terra sigillata, and smoke-fired. *Photo by Sumi von Dassow.*

Chapter 2

About burnishing

There are two methods of burnishing a pot: rubbing the clay with a polished stone or other smooth object, and coating the pot with terra sigillata and rubbing it with a soft material such as a chamois-leather. Using a stone is more time-consuming and takes a lot of practice, but can produce a higher degree of sheen. You also don't have to worry about the surface chipping or flaking off, and you can get a perfectly smooth surface with no brush-strokes or drip marks.

The most famous burnished pottery in the world is that of the Pueblo potters of the American Southwest. These potters achieve a brilliant shine on bare clay by rubbing it painstakingly with a polished stone before it is fired. Though burnishing this way is time-consuming and impractical for functional ware, the process is simple, and the result fascinating enough to make it worth any potter's time to experiment with it. The only materials needed are a suitable clay, water and oil; the only tool, a polished stone. The firing is to a low temperature, can be done in any type of kiln or even outdoors, and releases no toxic fumes.

Burnishing stoneware/ high-fire clay

Many types of clay can be burnished, though it should be smooth and grog-free. A cone 6 (not 06) clay will be the strongest and most durable after firing to cone 018 or so. (Most low-fire clays are so under-fired at such a low temperature that they can be scratched with a fingernail, like chalk.) Red clay seems to look more brilliant after polishing than white clay, and is often stronger both before and after firing. If a white clay is desired, choose a smooth cone 5 or 6 clay – not a porcelain or low-fire clay. White clays are best for bringing out the range of colour in a sawdust- or pit-firing. For the traditional all-black colour of southwest pottery a red clay may be preferable.

A suitable burnishing stone can be found at any lapidary shop, and often in museum gift shops or flea markets. Any kind of stone that has been tumbled in a rock polisher may be a perfect burnishing stone. An ideal stone is large enough to grasp easily, and has at least one perfectly smooth, slightly rounded surface. Any nick, bump, or sharp edge is likely to scratch the pot as it is being burnished: you can check a stone for nicks by running your fingernail over it.

For a first attempt at burnishing, start with a small rounded pot. Sharp angles, S-curves, flaring lips and grooves are difficult to burnish. To prepare the pot for burnishing, it should be sanded perfectly smooth when it is bone dry. If the surface isn't smooth enough, your stone will not be able to get into any little dips or depressions on the surface, and you will end up with dull, unburnished spots that the stone missed.

In addition to the pot and the stone, you need a towel or rag, a bowl of water and some vegetable oil. The first step is to wet the whole pot, inside and out, rubbing the water in quite thoroughly with your fingers. You need to rub the water into the clay before you start rubbing with the stone. This makes a bit of slip on the surface of the pot which helps to fill in any little scratches from the sandpaper, and dampens the pot so it doesn't dry out too quickly while you are burnishing it. Then you re-wet small patches with a finger, starting at the rim, and rub with the stone until the clay becomes smooth and takes on a dull sheen. The towel or rag is needed to wipe extra water off your finger and to wipe the stone clean as necessary. It is important to start at the rim, burnish the whole rim, then burnish all the way around just below the rim, continuing this way in a slightly overlapping spiral pattern from the rim towards the foot. Once a burnished patch has dried, you will scratch it if you rub it again with the stone. For this reason, once you begin you cannot stop until you've finished. Working in a spiral pattern ensures that you are never working on a patch adjacent to a section that has completely dried: by the time you are working on the bottom, the top – which may have been burnished an hour ago – is completely dry again, but the bottom ring of the spiral is still damp and can still be safely rubbed with the stone.

The second step is to cover the entire pot immediately with a light coating of vegetable oil and leave for 5–15 mins to soak in. After soaking in it leaves a whitish scum on the surface. Rubbed again with the stone, the clay takes on a high gloss. This step is much faster than

1 Wet the entire pot, inside and out, with your fingers dipped in water, then re-wet the inside of the rim. Work the water in well. With dry fingers, rub the stone on the dampened area, being sure to avoid leaving un-burnished streaks. If you have to rub hard to get the clay smooth, or if scratch-marks from sanding remain visible, you are not using enough water.

2 Once the inside rim has been burnished all the way around the pot, wet the top of the lip and burnish. The lip tends to be the most difficult area and should be carefully smoothed and rounded by sanding. As you burnish, bits of clay will collect on the surface you're working on. Use the back of your hand to wipe these bits off so that they don't cause scratches. Wipe clay off the stone with the towel or your thumb.

3 After burnishing all the way round just below the rim, continue in a spiral pattern towards the foot. When you dampen a new patch to burnish, slightly overlap the previously burnished area above. As long as an already-burnished spot has not changed colour (indicating it has dried completely) it is safe to go over it with the stone. As you progress down the body of the pot you can dampen fairly large areas and run the stone quickly over it horizontally and then more carefully vertically. Some people find it easier to burnish with a circular motion.

4 The last spot to be burnished is just above the foot (which is not burnished). Note the colour difference between the just-burnished area near the foot and the dry area at the rim.

5 The burnished surface is covered lightly with vegetable oil.

6 Holding the pot with a hand inside to avoid touching the surface, the pot is re-burnished wherever the oil has dried to a whitish scum. Burnish around patches of wet oil until they dry. If necessary, extra oil can be wiped off with a finger. If the stone scratches the oiled surface, add more oil and wait for it to soak in.

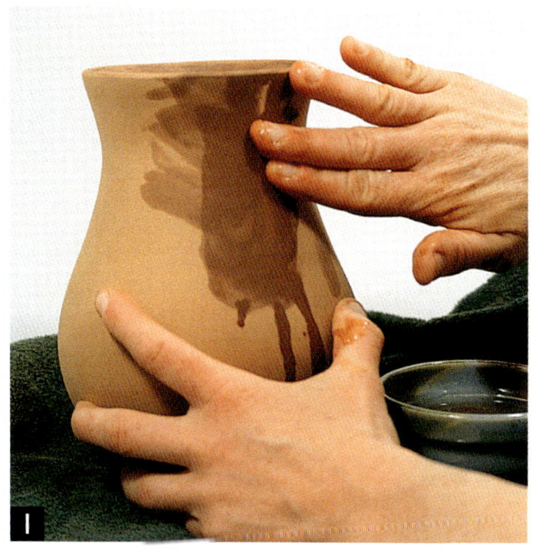
1

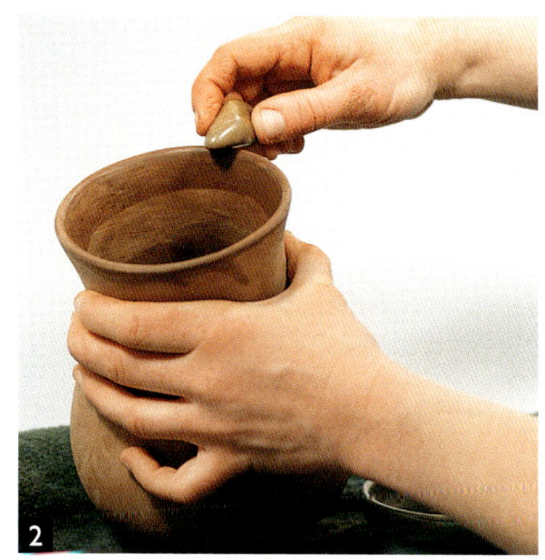
2

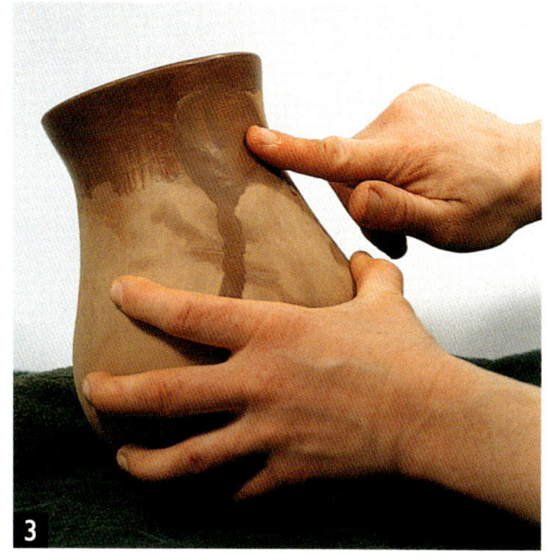
3

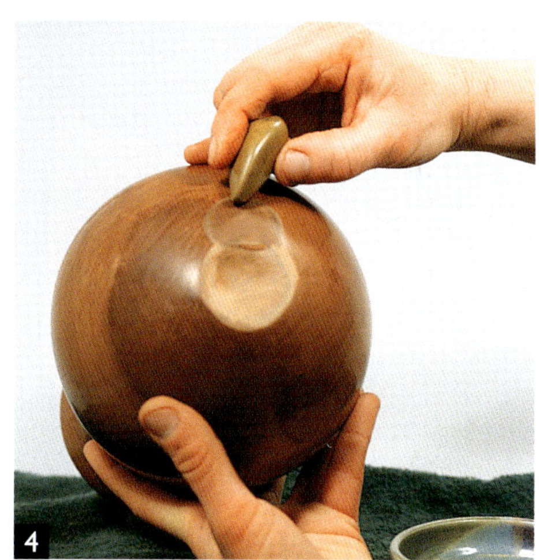
4

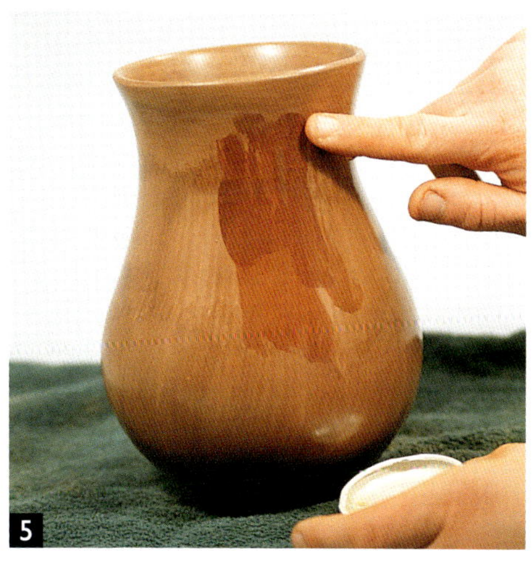
5

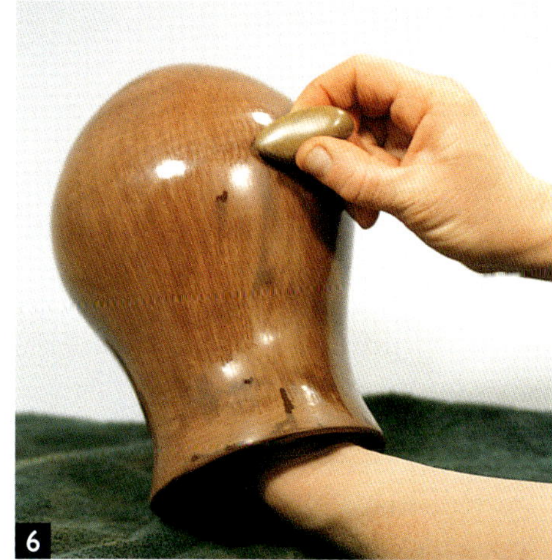
6

the first, and it doesn't matter what part of the pot is reburnished first. If there are patches of wet oil, you can work around them and reburnish the parts which are ready. During this step the pot will be easily marred by fingerprints and should be held with a hand inside. This will also help you avoid rubbing the oil off while you work. Lightly rubbing the pot with a chamois-leather after oiling and reburnishing will remove any extra oil and slightly improve the sheen.

The burnished pot can be decorated before firing by incising or by painting with terra sigillata or slip, or it can be immediately bisque-fired to cone 018. Though this is a very low temperature, it is high enough to harden the clay and drive off the internal water, without sacrificing the shine you worked so hard to achieve. Because of the very nature of clay, firing to a higher temperature dulls the burnish. Clay is made up of flat particles, called platelets. Burnishing works by pressing down the clay platelets on the surface of the pot, so they all face the same way and thus reflect light the same way. As clay is fired, it loses its platelet structure, so the higher the firing temperature, the more burnish you lose. With a suitable clay, cone 018 is hot enough to harden the clay without sacrificing too much sheen. Even if you are planning a subsequent firing such as a pit-firing, it is a good idea to bisque-fire your burnished pot to ensure that the finished pot is adequately durable, and to help the pot survive the fast temperature increase that is typical of many alternative firing methods.

Clays

A smooth (grog and sand-free) cone 6 red clay will have the highest fired strength. If a white colour is desired use a smooth, cone 6 clay. Navajo Wheel clay from Industrial Minerals Co. in Sacramento, California, is a suitable red body. B-Mix cone 5 from Laguna Clay Company is a good white body.

Burnishing a high-talc earthenware clay

Some potters prefer to burnish earthenware clay instead of stoneware. While earthenware may not be as durable when fired to a very low temperature, it can be polished to a remarkable level of sheen and the process is somewhat easier than with stoneware. Potter Michael Wisner has developed this technique to perfection, adapting methods learned from potter Juan Quezada of Mata Ortiz, Mexico.

To polish a high-talc body you need a polishing tool, baby oil and a damp rag such as a piece of T-shirt or stretchy cotton rag (something which will not leave bits of lint behind). The bone-dry pot should be sanded perfectly smooth, then covered with baby oil and allowed to rest for a few minutes while the oil sinks in. Then dip the piece of T-shirt cloth into water and wring it out so it is damp, not dripping, and then rub the pot all over with it. This helps to fill in small scratches from the sandpaper, and creates a light coat of slip on the surface. Then rub the pot all over with your chosen burnishing tool – Wisner burnishes with a stainless-steel car valve pushrod, instead of the traditional stone. There is no second step required, as you have already applied both oil and water.

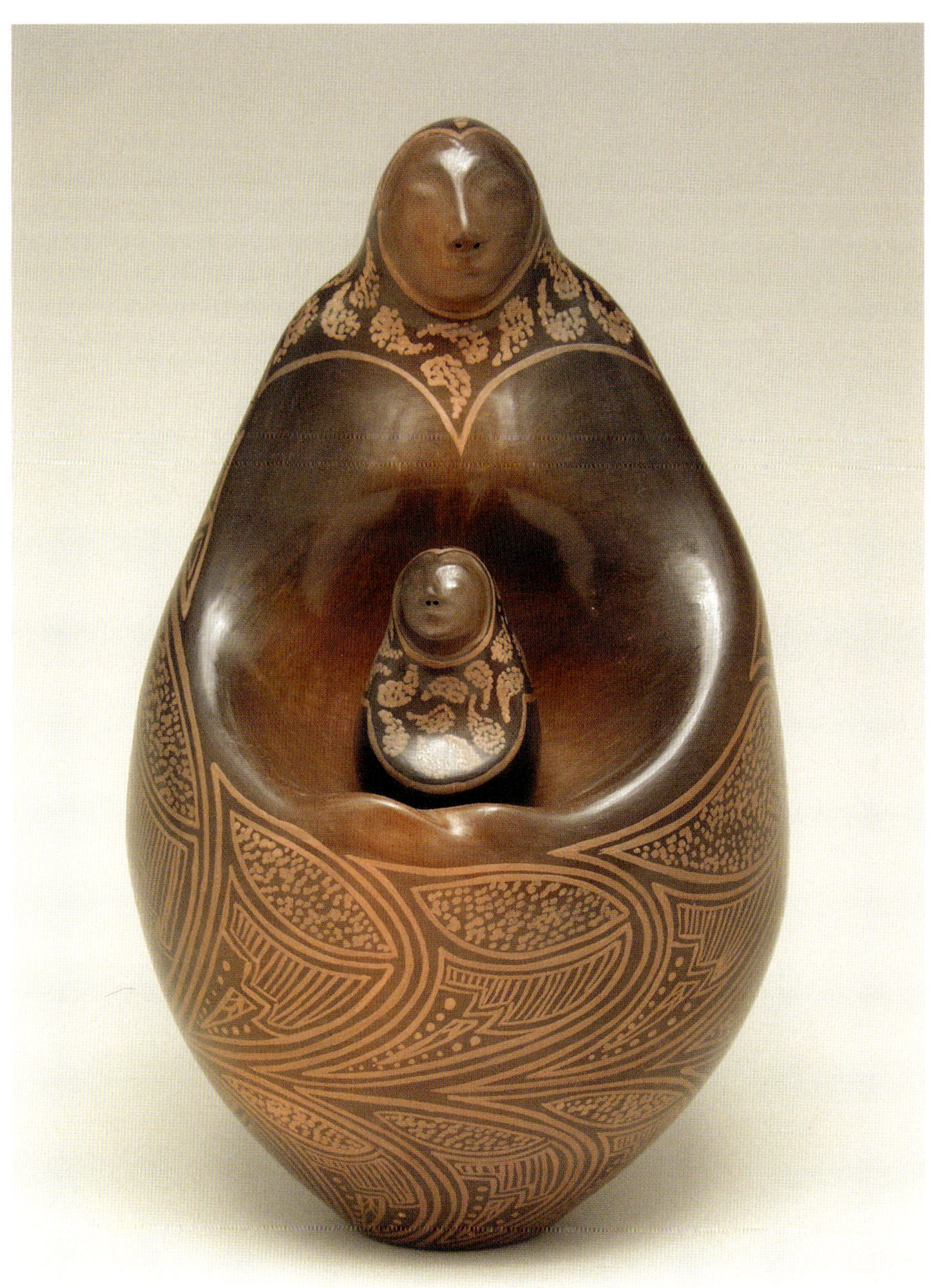

Mother and Child by Sumi von Dassow, 2004. 25 × 13 × 13 cm (10 × 5 × 5 in.). Coil-built from red stoneware (Navajo Wheel Clay from Industrial Minerals Co.), burnished with a stone, decoration painted on using terra sigillata and smoke-fired. *Photo by Sumi von Dassow.*

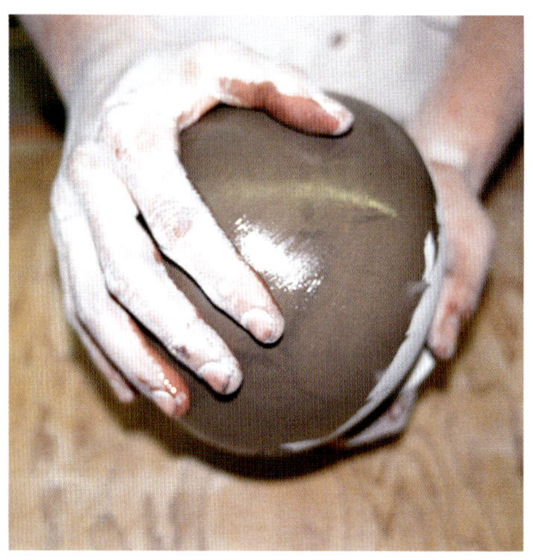

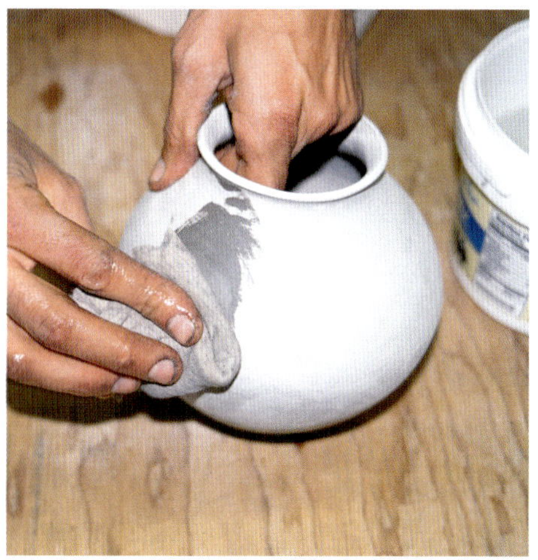

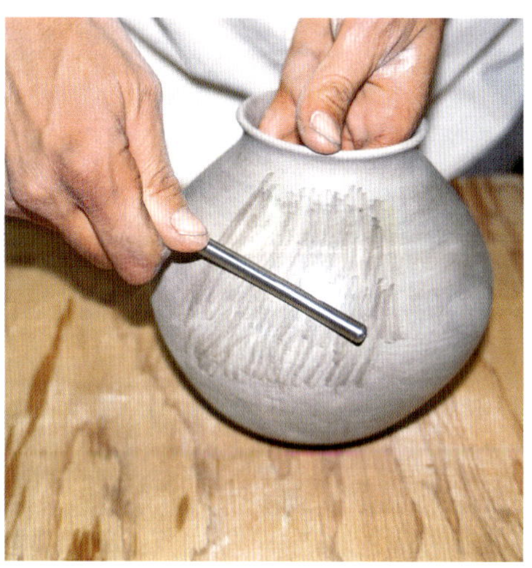

Some of the potters of Mata Ortiz use graphite in the burnishing process to achieve a metallic black surface. For this effect, Wisner mixes two tablespoons of powdered fine graphite with 125 ml (4 fl oz) charcoal lighter fluid. He prepares a pot for burnishing as above, then applies a thin coat of this graphite mixture with a disposable sponge brush, squeezing it out so the mixture doesn't run down the pot as he brushes it on. After the pot dries for a minute or two he rubs off any excess graphite with a piece of cloth, then burnishes it with the pushrod.

Pueblo Indian potters from the southwest USA dig their own clay, as do the Mata Ortiz potters. Each Pueblo has its own clay deposits and traditional methods of preparing clay and burnishing, but typically Pueblo potters make the pot out of clay from one deposit, and then coat the finished pot with a slip made from another clay before burnishing. Nancy Youngblood of the Santa Clara Pueblo burnishes her

LEFT, TOP TO BOTTOM
Michael Wisner coats a pot with baby oil.

Wisner rubs the pot all over with a damp piece of T-shirt cloth.

He uses a stainless-steel car valve pushrod to burnish the pot.

RIGHT, TOP Michael Wisner handbuilt vessels. Ht: up to 28 cm (11 in.). Burnished and decorated with slips, one with graphite.

RIGHT, BOTTOM *Red Melon Swirl* by Nancy Youngblood of Santa Clara Pueblo, 2000. Ht: 15 cm (6 in.), w: 20 cm (8 in.). Coil-built from clay collected on Pueblo land, carved, coated with red slip and burnished with a stone. *Photo by Hawthorne Studio, Santa Fe NM.*

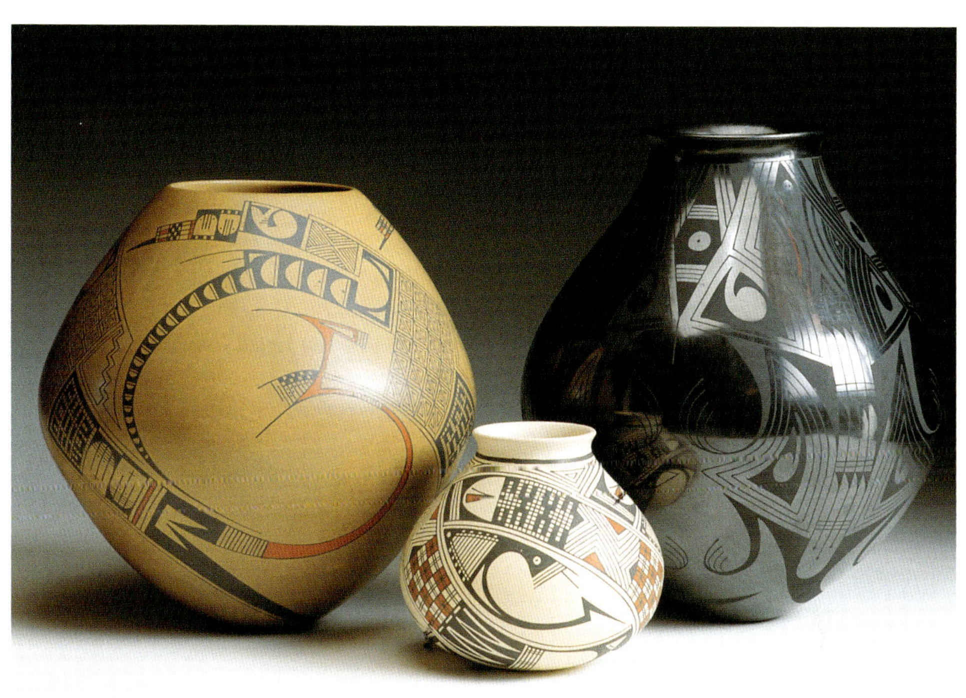

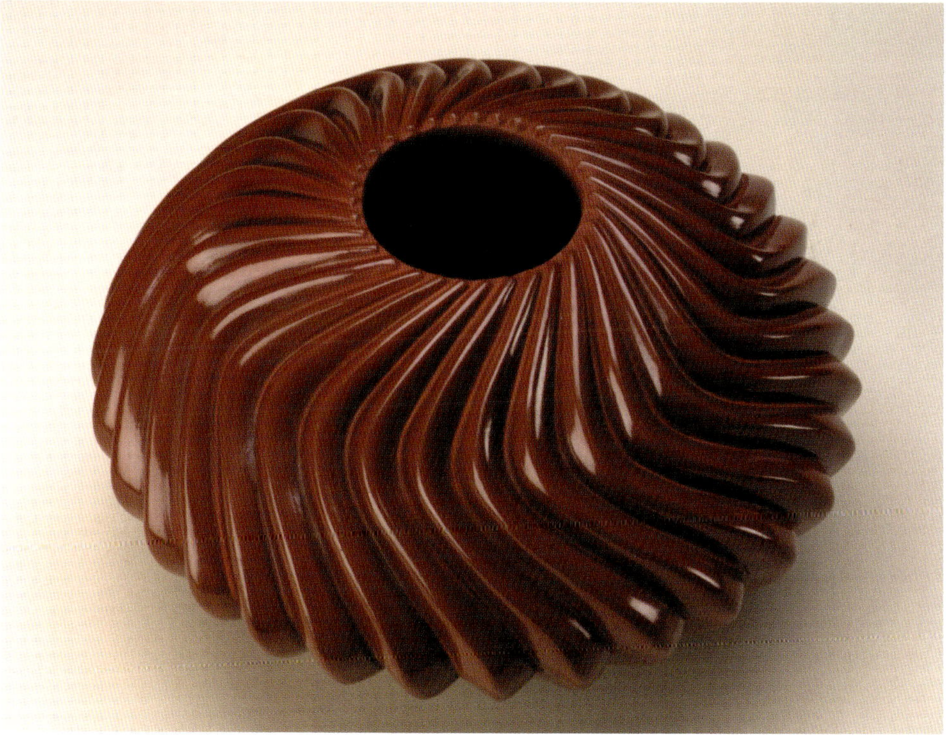

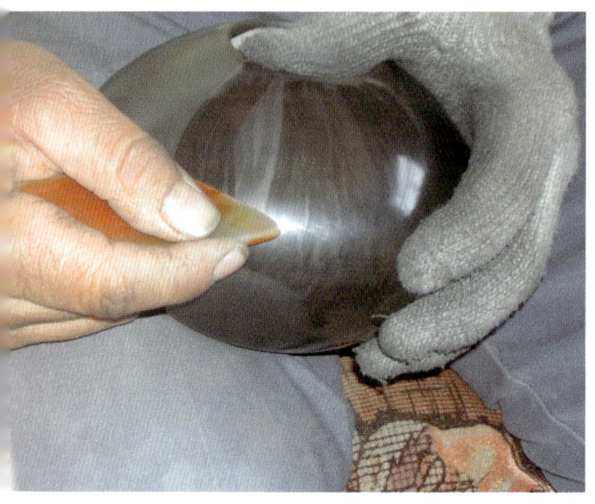

A Mata Ortiz potter uses a polished stone to burnish a pot that has been coated with a graphite mixture. *Photo by Ron Rivera.*

meticulously coil-built and carved melon pots in sections: on a melon pot with 64 ribs she burnishes three ribs in three hours, using a variety of burnishing stones handed down to her from her grandmother.

Clays

For a clay similar to that used by the potters of Mata Ortiz, Wisner uses CT-3 clay from Mile-Hi Ceramics in Denver, or his own formulation he calls 'Mike's Mud', composed of 50% Texas Talc and 50% Kentucky Ball Clay (OM-4) mixed with 2-5% bentonite.

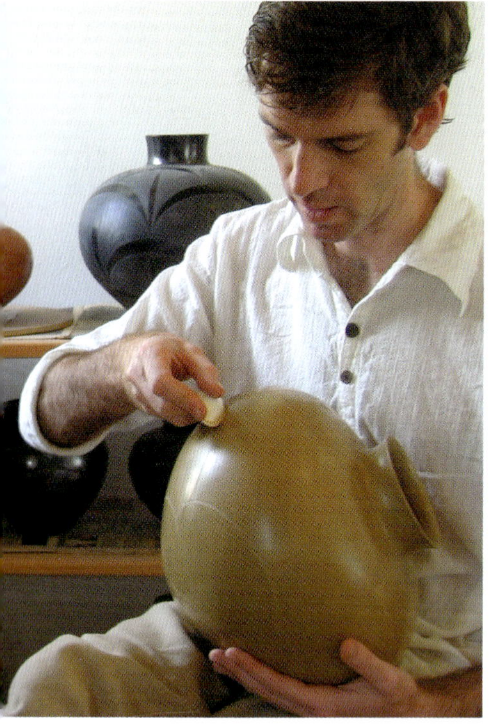

Ian Garrett burnishes his coil-built pots with a stone while they are leatherhard, then uses the serrated edge of a mussel shell to impress patterns into the surface.

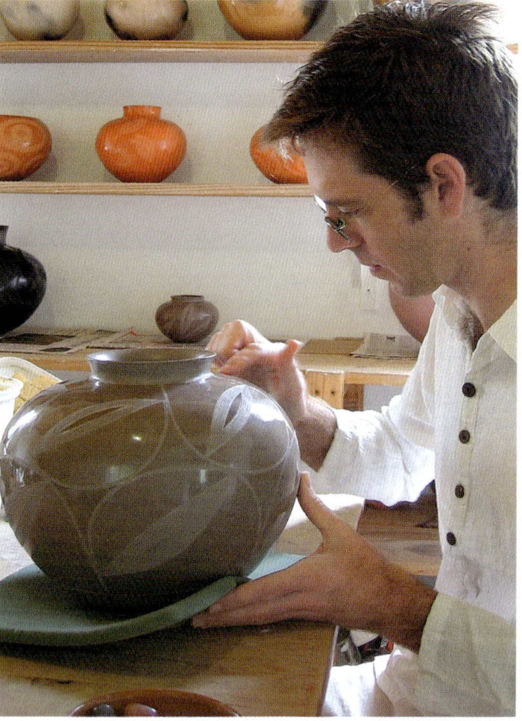

When the pot is bone dry, Garrett burnishes the entire pot to a high shine by lubricating small patches with vegetable oil and rubbing again with a stone.

Burnishing leatherhard or black-hard clay

Some potters find it easier to burnish a pot before it has dried completely. Timing can be tricky – you want the pot at the verge of dry, but with just enough moisture in the clay to allow your stone to glide across the clay without scratching it. Traditional leatherhard, the stage when a wheel-thrown pot can be easily trimmed, is a little too early. A leatherhard pot will show the marks of your burnishing tool as distinct little ridges, and in drying the rest of the way it will lose most of the shine you give it. Ideally, you want to catch the clay at black hard – when it is almost dry but

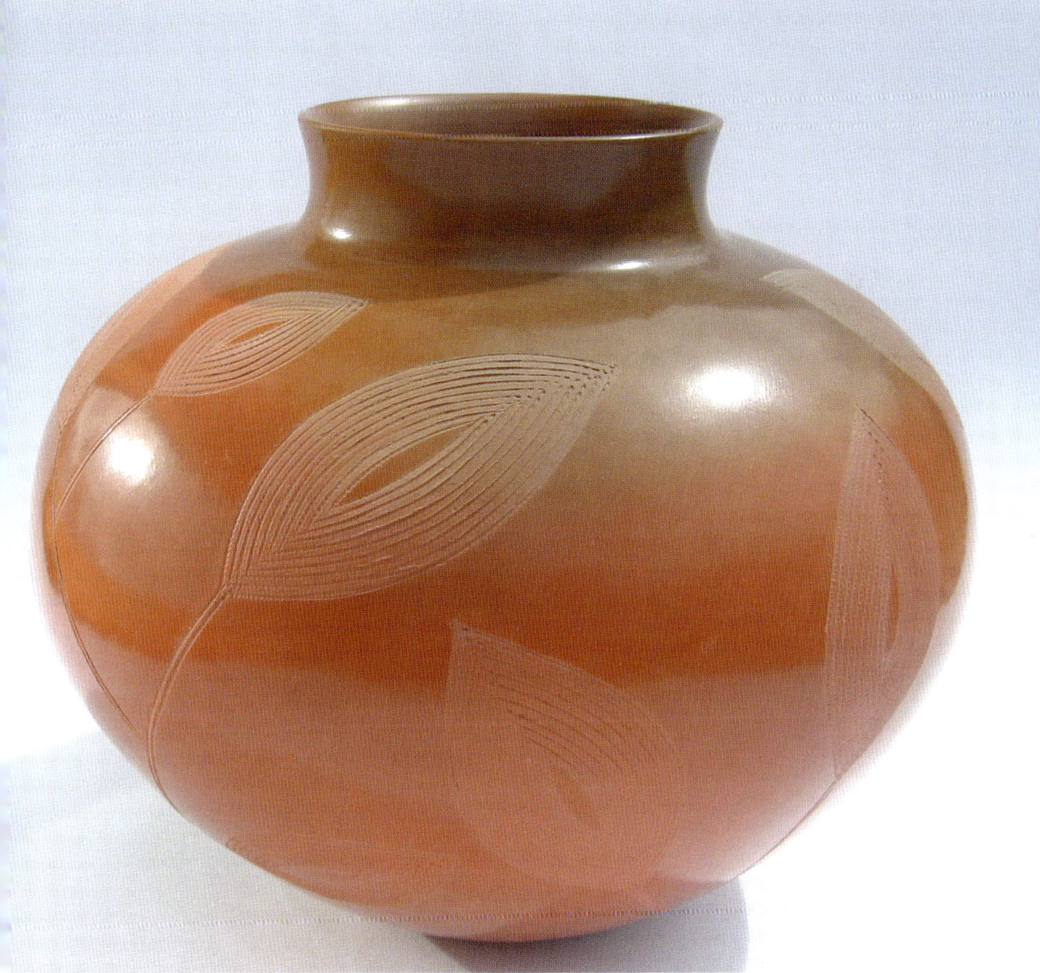

Autumn Leaves, by Ian Garrett, 2007. Ht: 17 cm (6½ in.). Decorated by impressing with mussel shells and burnished with a stone. Pit-fired.

has not yet changed colour. If you burnish at this point, then cover the pot to slowly dry the rest of the way, perhaps even going over it once or twice more with the stone before it dries completely, you can achieve a good polish. One drawback to burnishing this way is that you won't be able to sand the pot before burnishing, so this technique works better with a wheel-thrown pot that can be smoothed with a rib when wet, or after trimming. If you want to burnish this way, you have to pay close attention to the pot as it dries, checking and rechecking it, and burnishing and re-burnishing it until it is too dry to burnish without scratching.

In order to achieve the maximum level of burnish, Carol Molly Prier burnishes her pots four times, starting by first burnishing at leatherhard, immediately after trimming. If it is a handbuilt pot she scrapes it smooth at the leatherhard stage before burnishing. At this leatherhard stage she sometimes uses a flexible metal rib to burnish, instead of her stone. She burnishes twice more as the pot

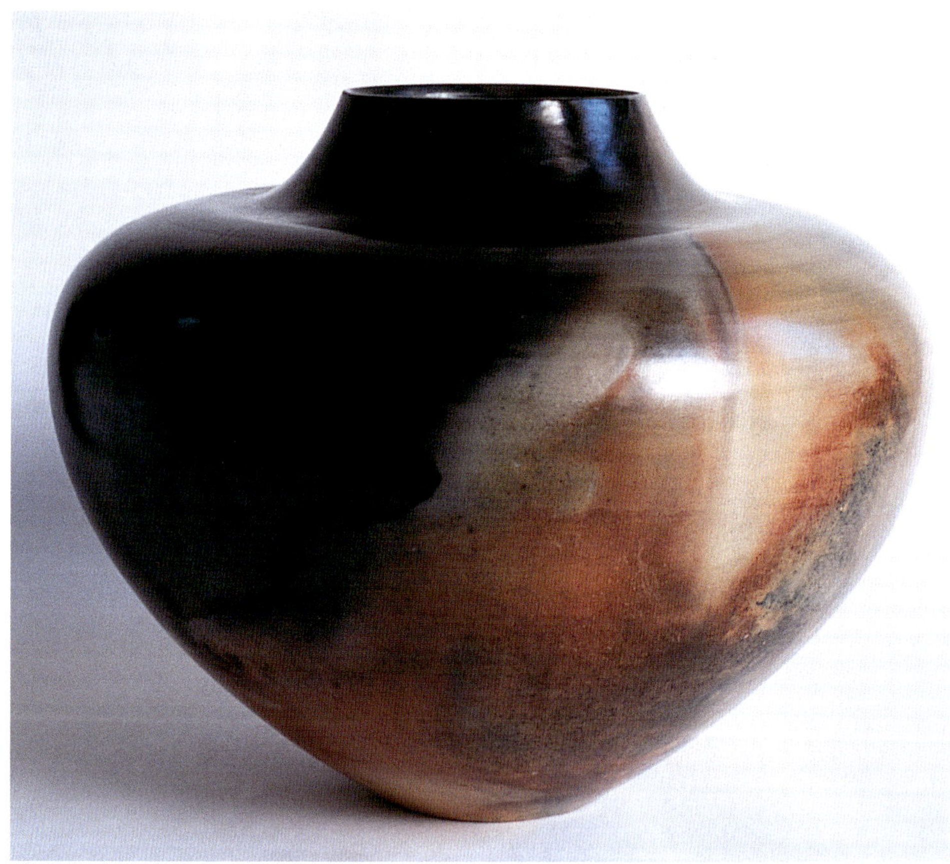

Pit-fired vessel by Carol Molly Prier, 2000. Dia: 30 cm (12 in.). Wheel-thrown and burnished.

continues drying, before it becomes bone dry.

Once the pot is bone dry she uses a soft facial tissue to apply a thin coat of salad oil over the entire surface of the pot. She lets it dry completely and then goes over the pot one last time with her stone.

Burnishing on the wheel

If you are burnishing a wheel-thrown pot, you may want to use the wheel to make the burnishing process easier. David Greenbaum burnishes wheel-thrown white earthenware pots on the wheel in a two-step process. The first burnish is when the pot is leatherhard, using a teflon plastic rib. When the pot is bone dry he rubs it all over with olive oil and allows it to soak into the clay. He then uses polished stones to burnish the pot again on the rotating wheel. Usually he makes three passes to eliminate any ridges the stone might have left the first and second time. With the pot still on the wheel he goes over the surface one last time with the Teflon rib to bring the

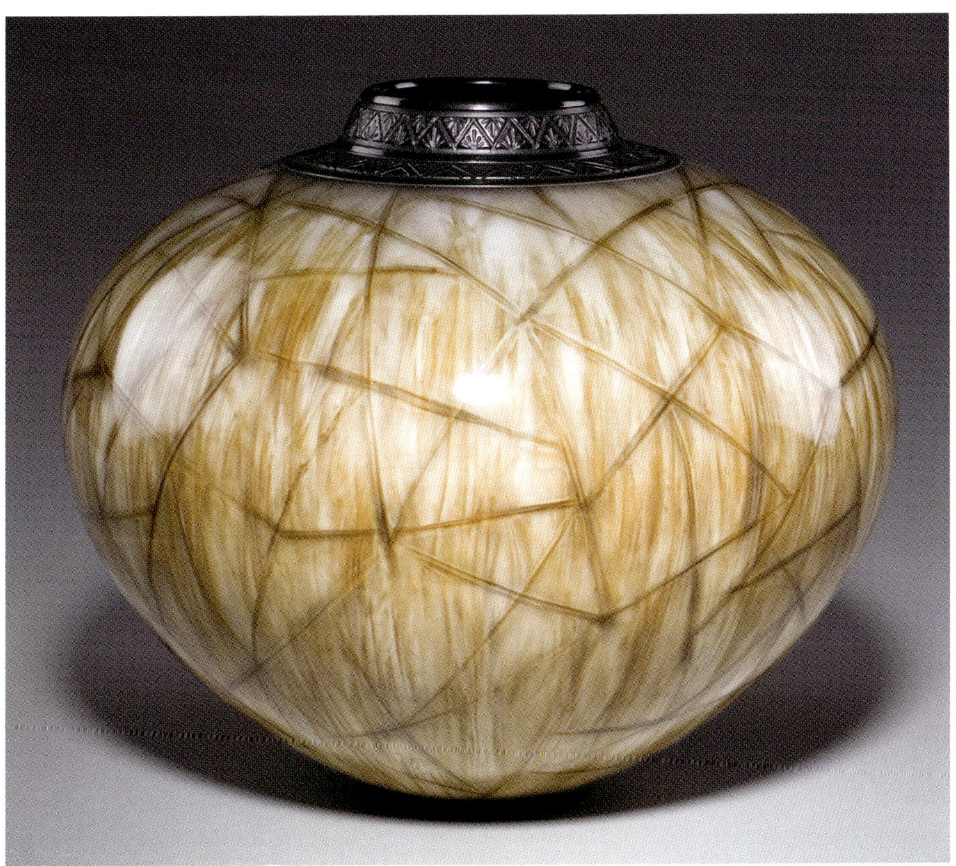

Latticed Globe Form by David Greenbaum. Dia: 40 cm (16 in.). Black-fired in an Axner Super Kiln in a saggar, partially buried in sawdust; then pit-fired wrapped with broom corn and grasses. *Photo by Randy Batista.*

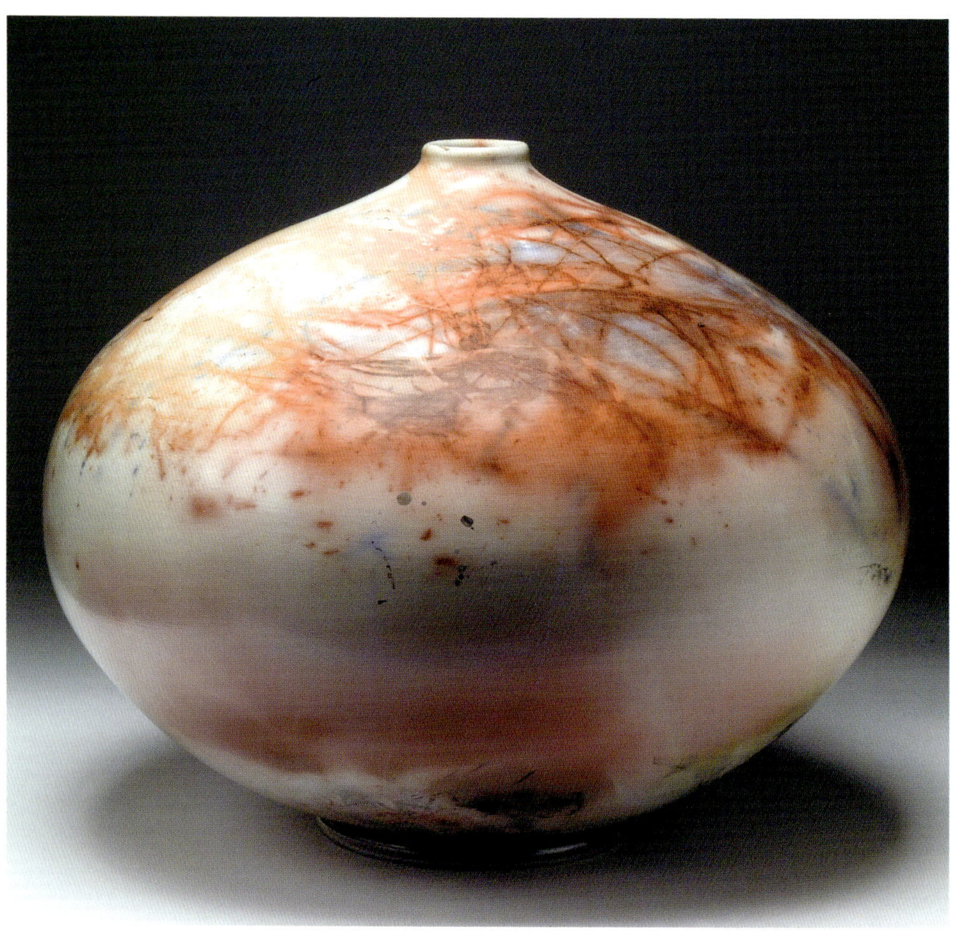

Saggar-fired vessel by Joan Carcia, 2007. 24 × 30 cm (9.5 × 12 in.). Coil built from low-fire white clay; burnished with a stone; saggar-fired with green hay, salt, seaweed, copper, salt marsh hay with iron oxide and cobalt carbonate. *Photo by Steven Gyurina.*

surface to a glass-like gloss. He uses a Giffin Grip to hold the pot on the wheel for burnishing – the small rubber 'hands' that hold the pot don't mar the surface as wads of clay might.

Burnishing tools

Any very smooth object can potentially be used as a burnishing stone. Many potters use rubber or plastic ribs for burnishing, particularly on leatherhard pots. The back of a spoon is a popular tool, though it may leave greyish marks on the clay. One of the more unusual burnishing tools I've heard of is used by Wally Asselberghs: he uses burnt-out lightbulbs of various sizes on leatherhard clay, because they are easier to grip. He does switch to a stone to finish the job once the pots are almost dry.

Clays

Greenbaum uses an earthenware clay called Miller 10 from Laguna Clay

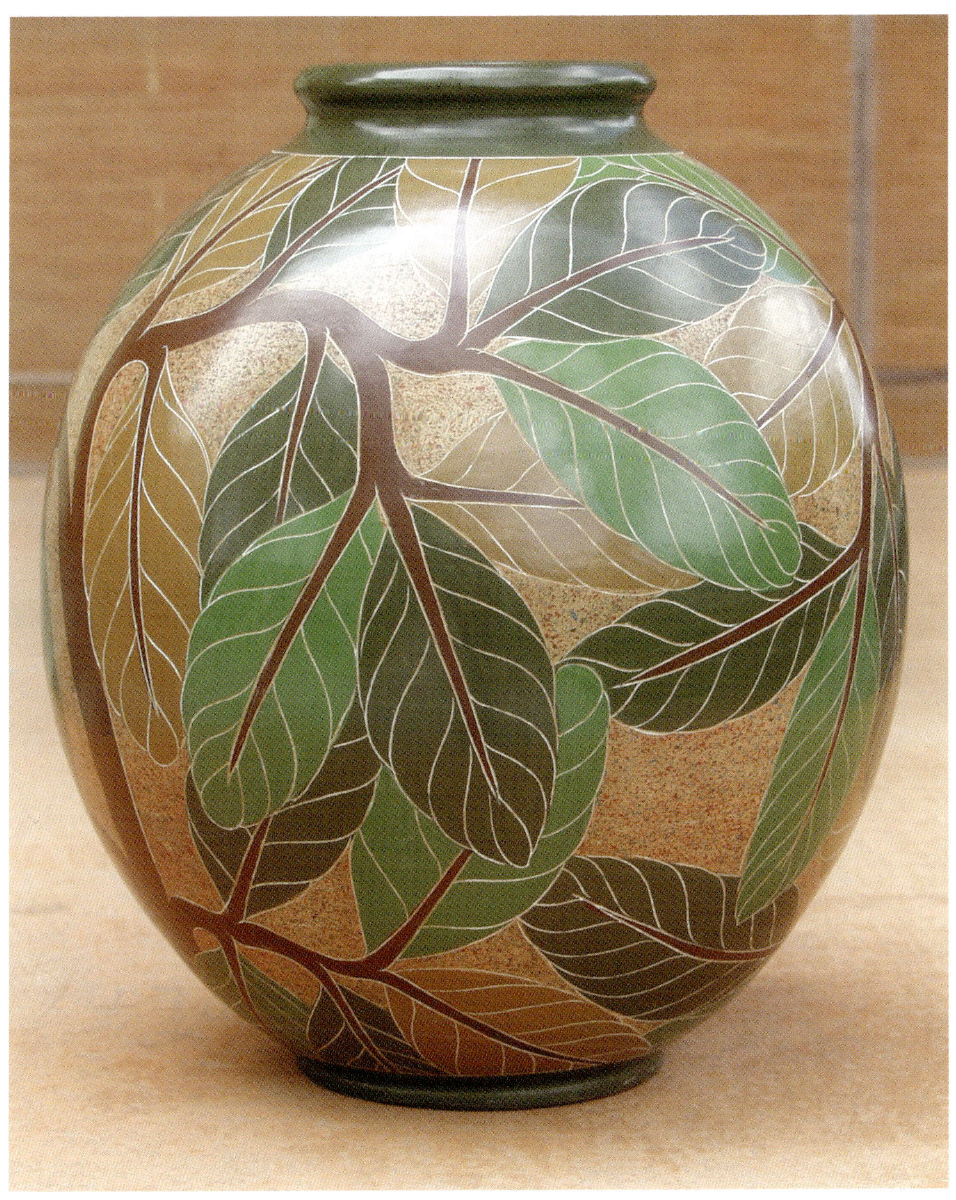

Vessel with Leaves by Blodimir Norori of Nicaragua. Ht: 30 cm (12 in.), dia: 20 cm (8 in.). With coloured slips and burnished in sections. Sgraffito lines delineate each coloured section to keep the colours from blending. *Photo by Sumi von Dassow.*

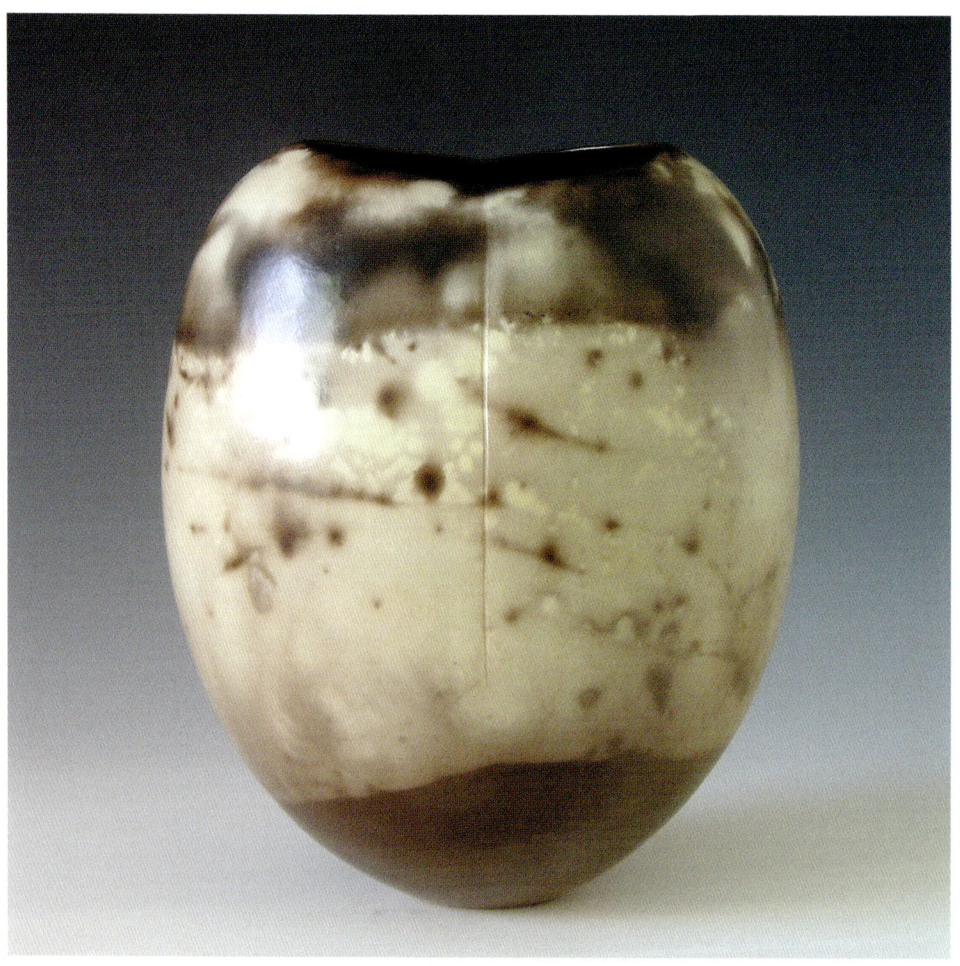

Pod Form by Gabriele Koch, 2008. Ht: 34 cm (13½ in.). Modified T-material, coiled, burnished slip, smoke-fired. *Photo by Kelpie.*

RIGHT *Desert Memory 15* by Carla Kappa, 2008. Ht: 71 cm (28 in.), dia: 28 cm (11 in.). Coil-built white earthenware (CT-3), burnished with pebble, fired to cone 04, with tree root.

Company, which is quite durable and not chalky, even when fired to a very low temperature.

Working in the UK, Gabriele Koch uses 'T' Material modified with White St Thomas Clay and CT1103 Porcelain, all from Ceramatech Ltd. Before burnishing, she coats her pieces with a screened slip made from the same clay body.

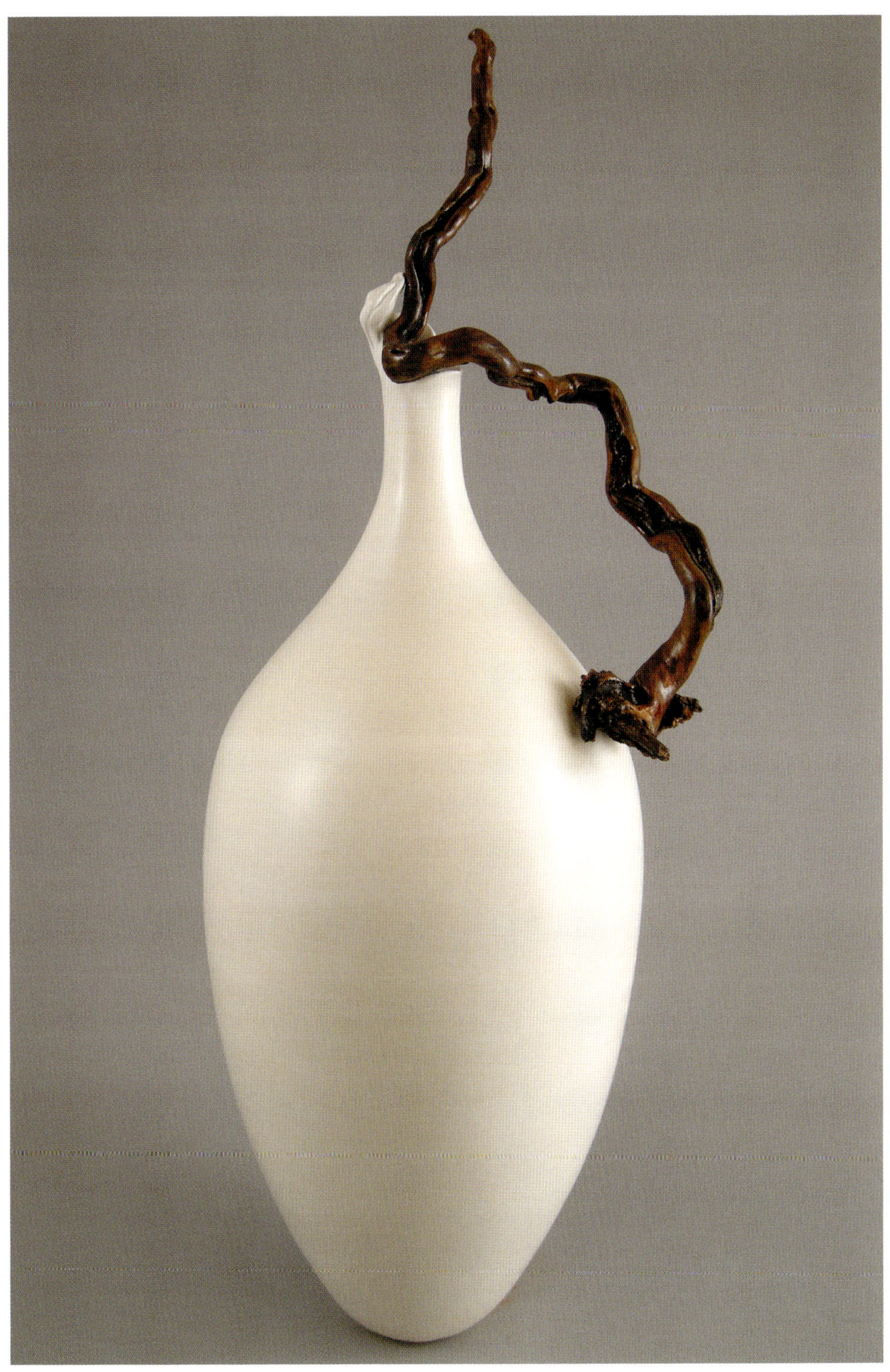

What kind of clay should I use?

Type of Clay	Burnishing	Black-firing	Pit or Saggar firing	Raku techniques
Porcelain	Can be burnished with a stone or use terra sigillata	Not recommended; may not get very black; finished work will be very fragile and may crack in fast firing	Not recommended; finished work will be very fragile, may crack in pit-firing, and some porcelains may not develop good colour. May be suitable for saggar firing.	Not at all recommended for raku. Likely to crack due to thermal shock.
Smooth White Stoneware *Example:* B-Mix cone 5 from Laguna, T-Material from Ceramatech, Little Loafer's Clay from Highwater Clay, Ash from Mile-Hi Ceramics	Smooth texture is ideal for burnishing with a stone or using terra sigillata	Not recommended: may not get very black; will not be very strong; and may crack in fast firing	May be very suitable. Some white stonewares work very well for these techniques. May pick up better colour with terra sigillata than burnished with a stone.	Not ideal; may be likely to crack due to temperature shock. Many potters like the smooth texture and are willing to risk some pots cracking.
Smooth Red (high-iron) Stoneware *Example:* Navajo Wheel from IMCO, Redstone or Earthen Red from Highwater Clay	Smooth texture is ideal for burnishing with a stone or using terra sigillata. Red clay is especially beautiful when burnished with a stone.	May pick up rich black colour when burnished with a stone and bisqued to low temperature (cone 018)	May be very suitable. Red clay can be beautiful fired with these techniques, but range of colour will be smaller.	Not recommended; may be likely to crack due to temperature shock.
Groggy stoneware, raku clay or sculpture clay. Many clays from every manufacturer fall into this category	Difficult to burnish with a stone due to coarse texture. Can use terra sigillata but coarse texture will show.	May turn black, but black-firing looks most striking on a stone-polished pot so groggy clay is not recommended.	May be very suitable. Groggy clays will resist thermal shock due to fast firing; a slightly groggy buff stoneware will pick up good colour either burnished or with terra sigillata.	Ideal for fast-fire raku techniques, but some potters prefer smoother clays.

Type of Clay:	Burnishing	Black-firing	Pit or Saggar firing	Raku techniques
Paperclay *Example:* Biz Bod P'Clay (high fire) or CT-3 P'Clay (low-fire) from Mile-Hi Ceramics – or make your own	Can be burnished with a stone or use terra sigillata. Difficult to get as highly polished as a smooth clay body.	May be suitable for some black-firing techniques, though black-firing looks most striking on a highly polished clay.	Very suitable, especially for pit-firing. Paperclay is highly resistant to thermal shock. May pick up better colour with terra sigillata than burnished with a stone	Very suitable. Paperclay is highly resistant to thermal shock and will have a smoother surface than raku clay.
High-Talc Earthenware *Example:* CT-3 from Mile-Hi Ceramics, several clays from Laguna (but not Miller 10) White Earthenware from Highwater Clay	Can be burnished to a high gloss with a stone using baby oil technique, or use terra sigillata	May be very suitable as black-firing looks most striking on a highly polished surface. Some earthenwares will pick up black better than others.	May be suitable, may pick up nice colours, but many earthenwares are very fragile when bisque-fired to low temperatures.	Not recommended; can be prone to cracking due to thermal shock. But if you like the way your earthenware burnishes, try it.

Note: Most clay manufacturers make clays in all the above categories. Your supplier will be able to tell you if a clay is smooth or groggy, and if an earthenware is high or low in talc. Though specific clay bodies are recommended in this book, many different clay bodies can be used for these techniques. Try whatever clay you enjoy working with, or ask for a small sample of any clay that sounds suitable from your local supplier. Most suppliers will give you small samples free, or for little charge. Experiment with several clays and several burnishing and firing techniques. Once you find a clay body that responds well to burnishing and firing, and that you enjoy working with, you will be well on your way to developing your own unique style of burnished work.

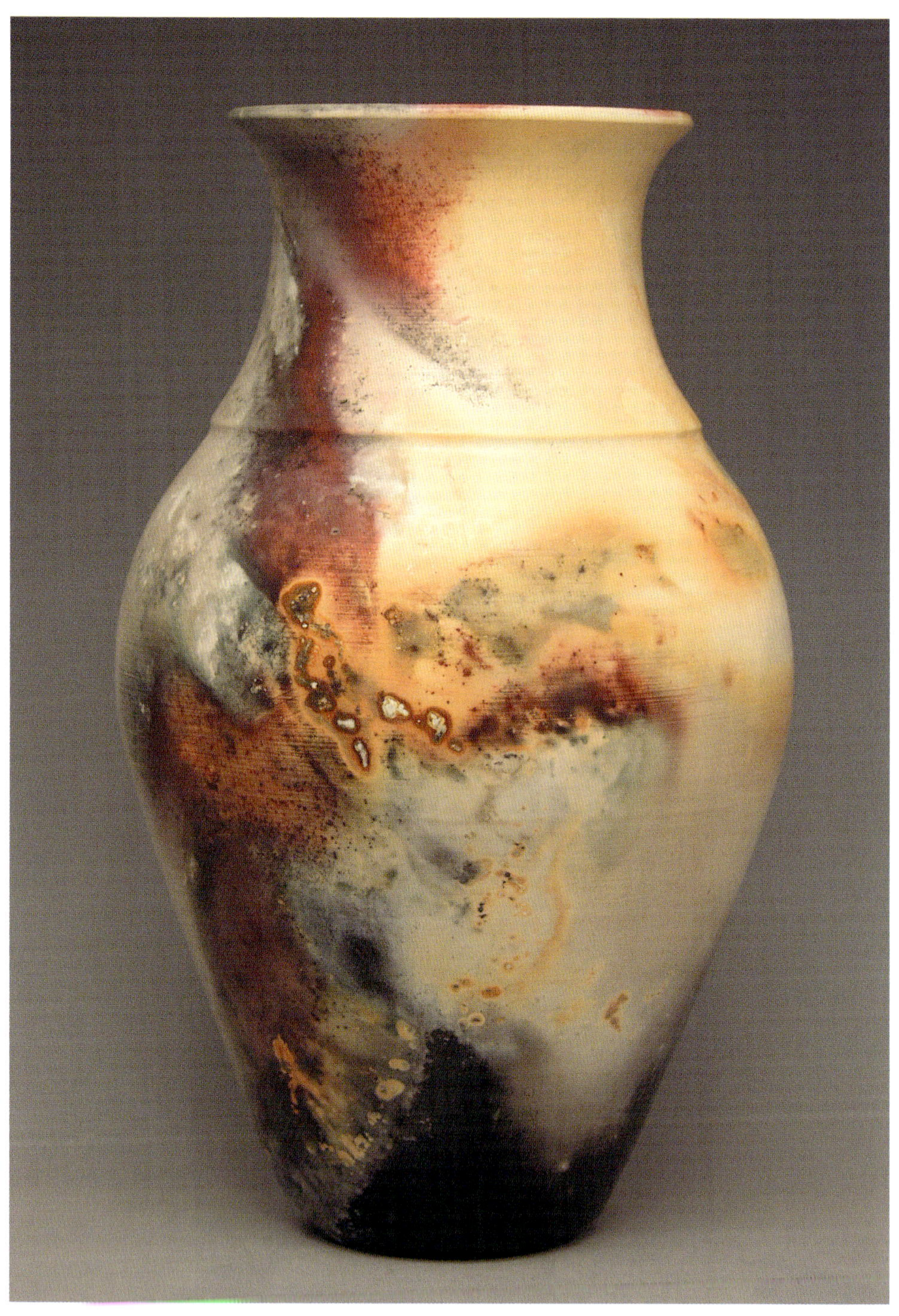

Wildfire by Sumi von Dassow, 2007. Ht: 30 cm (12 in.), dia: 15 cm (6 in.). Pit-fired pot with brushed terra sigillata. Horizontal traces of brush strokes are evident if you look closely.

Chapter 3
Burnishing with terra sigillata

Some potters think of terra sigillata as the easy way to burnish. Terra sigillata is relatively quick and easy to apply, but there are various potential disadvantages compared to stone polishing: it can crack or peel off after firing; it may be difficult to apply an absolutely smooth coat without brush marks or drips; it will reveal all texture on the surface it is being applied to, for good or ill; and it will never quite yield the mirror-like shine you can achieve by burnishing with a stone.

Which method of burnishing you choose will depend on the clay you are using, the way you are firing and your level of patience. The form you are trying to burnish may also dictate which method you can use – a smooth surface with no folds or creases, and no impressed, incised or sculpted details is ideal for burnishing with a stone. If your form has a textured surface, or corners and curves which a stone cannot get into, then you probably need to use terra sigillata.

The term terra sigillata, which means 'sealed earth', comes from the name of a type of Roman pottery mass-produced around the first century AD. This pottery was decorated with impressed or stamped decoration, which is what the word 'sigillata' refers to. (Think of the kind of stamp, or 'seal', which would have been used to seal wax on a paper document.) These pots were coated with the same kind of very fine slip which Greek potters had been using for hundreds of years to create their famous black and red pottery. Though many books incorrectly refer to this slip as a 'glaze', it was not actually a glaze but the material we now call terra sigillata.

Making terra sigillata

Terra sigillata, or 'terra sig' for short, is made by mixing a suitable clay with water and a deflocculant and leaving it to stand until the heavier particles of clay settle out. (Deflocculant weakens the electrical attraction between particles of clay, thus breaking up small clumps of clay and allowing the individual particles to float freely.) The deflocculant causes the finer particles to float in the water, which can then be decanted for use. In general it is not possible to buy terra sigillata, so if you want to use it, you must make your own.

To make terra sigillata, you will need a clear glass or plastic jar with a wide mouth, an accurate gram scale and a length of clear plastic tubing for siphoning. The only ingredients are water, dry clay and deflocculant. Many kinds and colours of clay can be used,

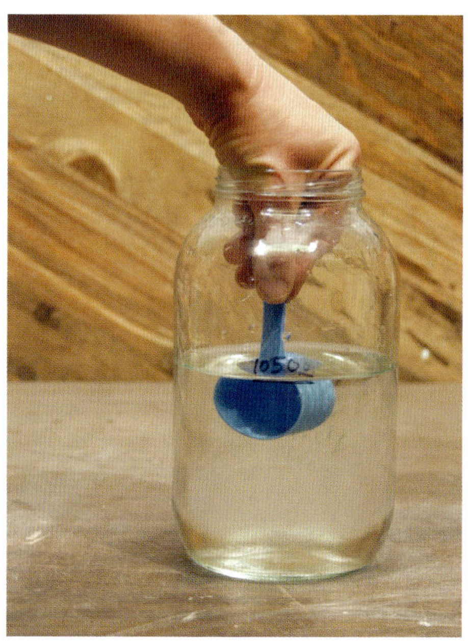

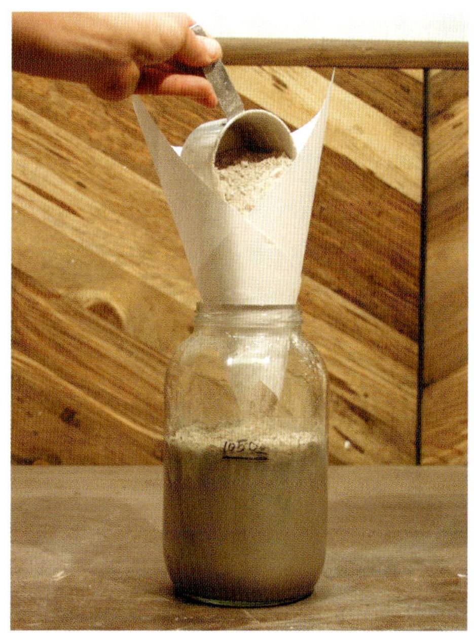

Step one, (*left*): Weigh out Darvan 7 or 811 and add it to the measured amount of water, making sure to get all the weighed material into the water. *Helpful hint:* If you always mix your terra sigillata in the same container, place a permanent water-line on the side of the container so you don't have to measure the water every time.

Step two (*right*): Weigh out the dry clay and add it to the water. Stir or shake thoroughly.

including ball clay, kaolin, local clay or scraps of whatever clay body you usually work with. There are also many possible deflocculants, the most commonly used being sodium silicate, soda ash, Darvan 7 and Darvan 811. You might find recipes calling for Calgon water softener, but don't try those – unfortunately Calgon doesn't work since it was reformulated to eliminate phosphates. Lye can also be used as a deflocculant, and I have even experimented with using the waste water from washing wood ash.

Not all clays are equally suitable to make terra sig, and the proportions of water to clay to deflocculant will be different depending on what clay you use. It is a question of experimenting with different types until you find something suitable. You can try substituting any dry clay, including scraps from your clay body, for the clays called for in the recipes which follow (see p.42). The process is simple, but a bit time-consuming. First, measure your water, and stir in the deflocculant. Weigh out your clay and add it to the water. For best results, be sure to weigh these materials precisely. If you have a ball mill, you can ball mill the mixture; otherwise, shake or stir it vigorously. Then place the jar, loosely covered, somewhere where it won't be disturbed for several hours to several days, depending on the recipe.

After the appropriate settling period, you will see a layer of dark sludge on the bottom of your jar, and if it has been a

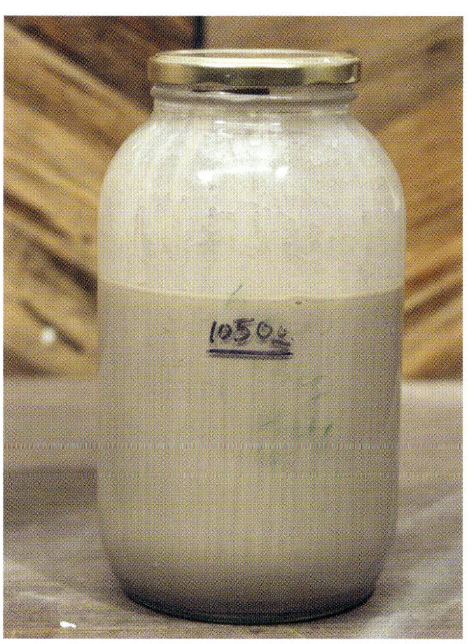

This shows stripes down the side of the container as the heavier particles begin to settle.

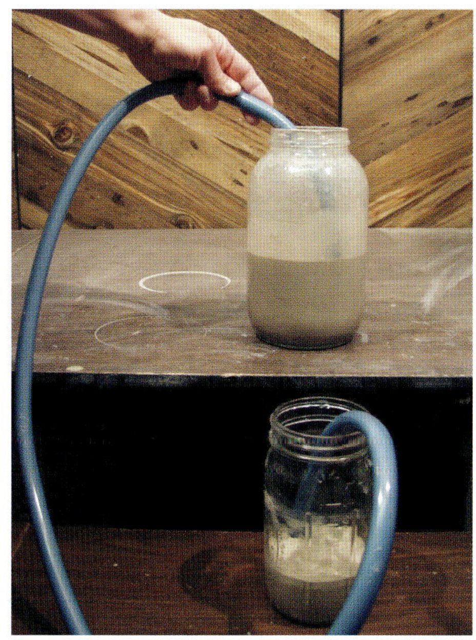

After settling (in this case, approx. 3 hours), siphon off the liquid layer from above the layer of sludge. Be careful not to get any of the settled sludge in your siphon!

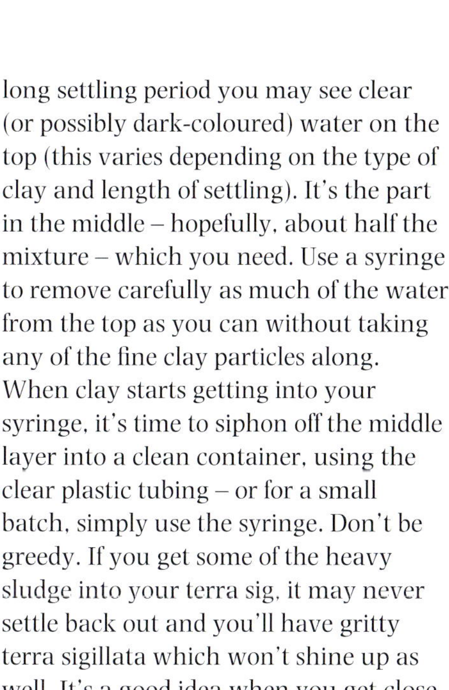

long settling period you may see clear (or possibly dark-coloured) water on the top (this varies depending on the type of clay and length of settling). It's the part in the middle – hopefully, about half the mixture – which you need. Use a syringe to remove carefully as much of the water from the top as you can without taking any of the fine clay particles along. When clay starts getting into your syringe, it's time to siphon off the middle layer into a clean container, using the clear plastic tubing – or for a small batch, simply use the syringe. Don't be greedy. If you get some of the heavy sludge into your terra sig, it may never settle back out and you'll have gritty terra sigillata which won't shine up as well. It's a good idea when you get close to the layer of sludge to switch to a new container, so if some of the sludge gets in you don't contaminate the whole batch. You now have a batch of rather thin terra sigillata (along with a lot of sludge which can be discarded or used for some other purpose). It can be used as is, or allowed to settle and evaporate for a few days before using.

Applying terra sigillata

Terra sigillata can be applied in two ways, by brushing or spraying. Brushing is easier, but may leave noticeable brush-marks. On the other hand, spraying requires more equipment, and may leave a bit of a pebbly texture where the droplets land on the pot.

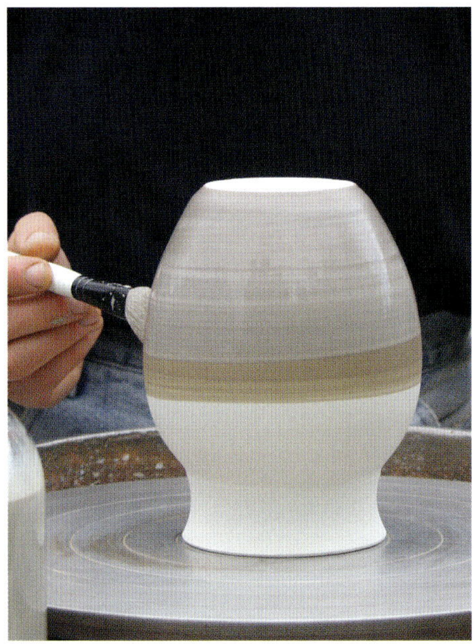

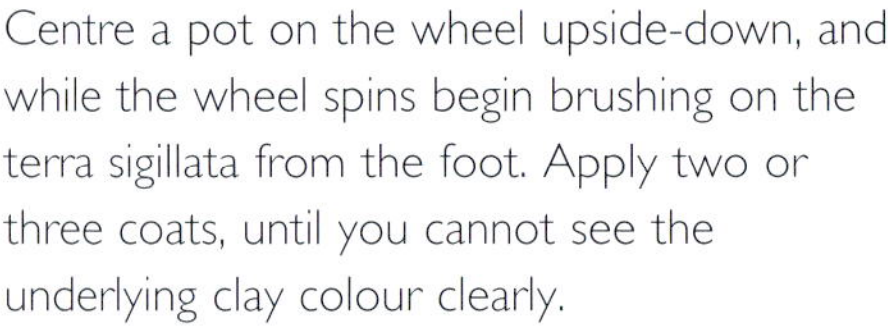

Centre a pot on the wheel upside-down, and while the wheel spins begin brushing on the terra sigillata from the foot. Apply two or three coats, until you cannot see the underlying clay colour clearly.

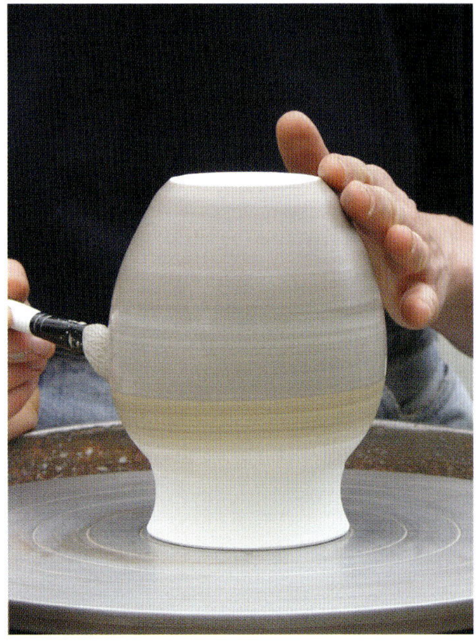

Once you have applied enough terra sigillata to the lower portion of the pot, and it has soaked in so it is not glossy wet, use your fingertips to begin polishing the surface as you continue coating the pot with the other hand.

Before you apply terra sigillata, your pot must be smooth and dust-free. Terra sigillata is so fine that even if you cover a textured surface with several coats, the texture still shows. This is wonderful if you have a deliberately textured surface, and in fact, the only way to burnish a textured surface is with terra sig. However, if you have sanded the pot, the surface is likely to be covered with little scratches from the sandpaper, which will not be covered up by the terra sig. Even more important, if you have sanded your pot, you must carefully sponge off any dust. Dust will cause the terra sigillata to peel off after firing. Therefore, if you want to achieve a really smooth burnished surface using terra sig, it is most effective to apply it to a wheel-thrown pot which has been ribbed smooth after throwing or trimming, or if handbuilding, to smooth the entire surface at leatherhard stage with a rib.

Terra sigillata should be applied to a bone-dry or almost bone-dry pot. If you are brushing it on, you need to apply at least three coats. If you are putting white terra sigillata on white clay, three coats is probably plenty. The terra sigillata needs to soak into the clay, but should not be allowed to dry completely between coats. Once you have applied several coats, the surface should be buffed with your fingers, a cloth or chamois-leather before it dries completely. The pot is ready to buff when the surface looks waxy and grey but is no longer wet-looking. If it has lightened in colour, it has dried too

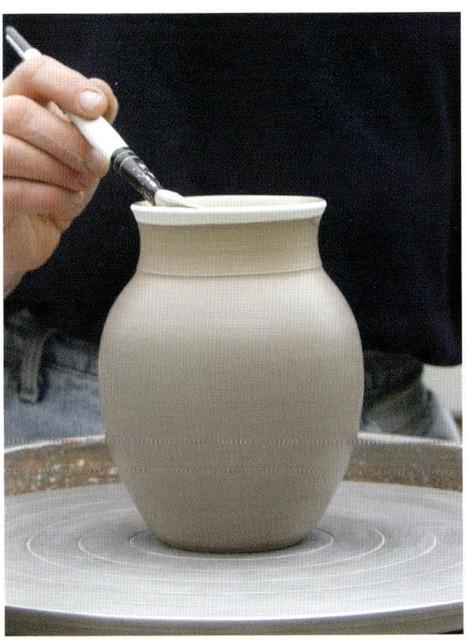

Turn the pot right-side up and finish coating it with terra sigillata. Apply just inside the lip; there is no need to coat the entire inside.

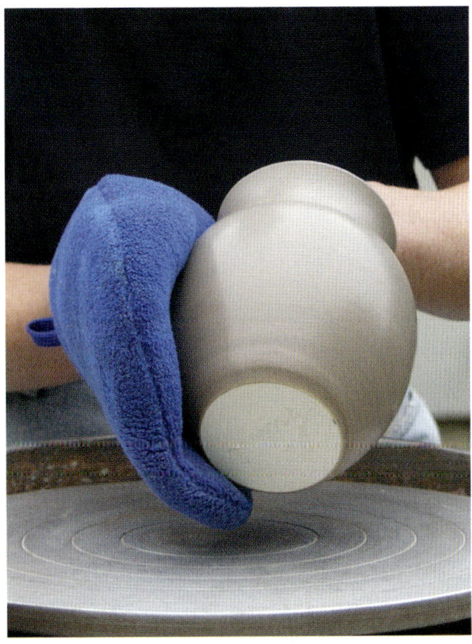

Using the polishing material of your choice – a car-polishing mitt works well – bring the entire surface to a high sheen.

much and another coat of terra sigillata must be applied. For the greatest degree of sheen, apply three thin coats and buff after each coat.

Watch out for two things when you are brushing on terra sigillata: don't let it drip down your pot, because the drips will show; and don't allow your brush to lose hairs, as the hairs will make a permanent mark. Be sure to use a good-quality soft brush – a watercolour mop brush works well. If you are brushing terra sigillata onto a wheel-thrown pot, the simplest way to apply a nice even coat is to put the pot on the wheel and let the wheel do the work for you while you move the brush up and down. Once you have enough coats on part of the pot, you can start burnishing with the fingertips of one hand while you are still brushing the terra sigillata onto another part of the pot with the other hand. If you have a large pot you are almost required to do this to get a really good polish, or the terra sigillata may dry out too much before you finish brushing it on. Don't touch the surface until it has soaked in, though if the terra sigillata comes off on your fingers, it isn't ready to burnish yet, and you will mar the surface by touching it. After you have applied enough terra sigillata to the whole pot, and there are no wet patches, then you can start using a chamois-leather or a soft cloth, or even a thin plastic shopping bag, to bring the surface to a high gloss.

If you are applying terra sigillata to a handbuilt or sculptural piece, you may

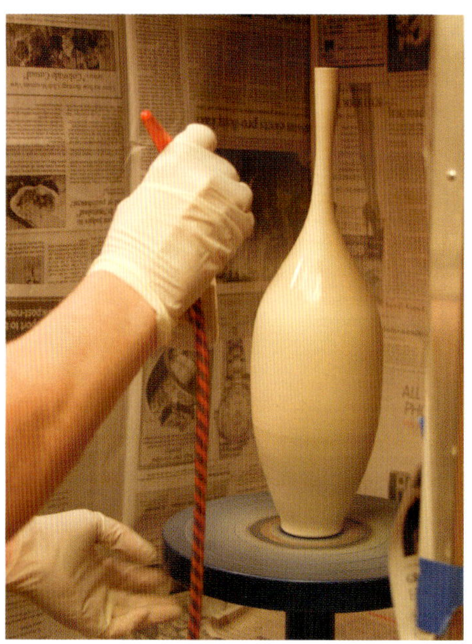

Linda Dadisman sprays terra sigillata using an airbrush. Be sure to do this in a properly vented spray booth, and wear a respirator. She wears rubber gloves to avoid marking the pot with fingerprints.

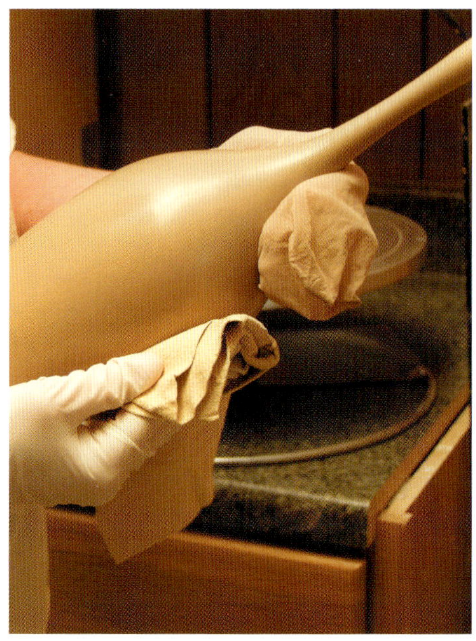

After two or three passes with the airbrush, Linda stops and polishes with a nylon stocking. After two or three more passes with the airbrush, Linda does a final polish with a chamois-leather.

find it impossible to use the wheel to help with the job. In that case, you can still brush it on, but be careful not to touch any wet spots. You may also want to experiment with pouring, or even dipping if you have a large enough batch of terra sigillata and a way to safely hold a delicate piece of greenware.

Some potters prefer to spray on terra sigillata for a more even coat with no brush-strokes. You can use a glaze spray gun or an airbrush, but of course the airbrush will produce a finer spray. When spraying, you must take care not to hold the sprayer too close to the pot or the wet spray will collect and drip – but if you hold the sprayer too far away, the spray will dry too much before it lands and create a slightly pebbly texture. It will help to have the pressure fairly low, no more than 13.5 kg (30 lb). Once you have applied enough coats to create a waxy surface which doesn't dry too quickly, put down the spray gun and buff it with your fingers, a cloth or a chamois-leather. You may want to buff once or twice between coats, also. Be careful not to touch the surface of the pot while the terra sigillata is wet, and make sure your cloth or chamois-leather is very clean so it doesn't scratch the surface.

RIGHT Saggar-fired bottle form by Linda Dadisman, 2007. Ht: 25 cm (10 in.), W:10 cm (4 in.). Sprayed with terra sigillata. Saggar-fired with salt, copper carbonate, steel wool and pet bedding. *Photo by Shelly Schreiber.*

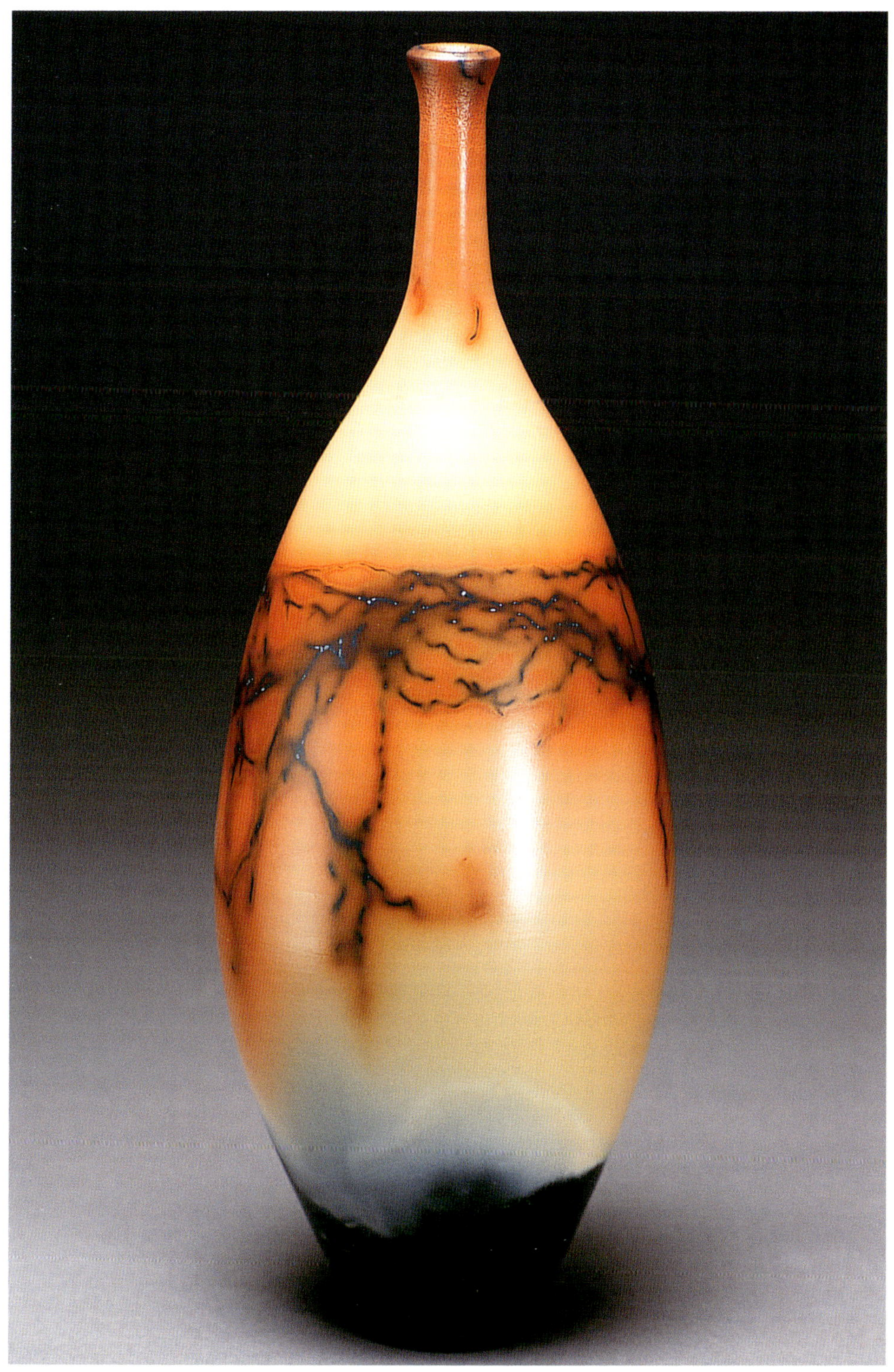

Recipes

White terra sigillata

2.1 kg (2.1 litres/3.7 pints) water
1 kg (2.2 lb) OM4 Ball Clay
25 g Darvan 7 or Darvan 811

Red terra sigillata

2.2 kg (2.2 litres/3.9 pints) water
1 kg (2.2 lb) Newman Red Clay
30 g Darvan 7 or 811

- Measure water into a large glass or clear plastic jar with a wide mouth.
- Add Darvan and stir. Add clay and shake vigorously. Leave undisturbed to settle for three hours.
- You will see a dark layer of sludge at the bottom of the container. Siphon off the liquid portion above the layer of sludge. Be careful you don't pick up any of the sludge. Discard the bottom layer of sludge.
- Try the terra sigillata on a dry test tile. If it doesn't give an adequate shine or still feels gritty when you rub it with your fingers, let it sit for another 12 – 24 hours, siphon again and discard the new sludge. One sign that you have a bad batch of terra sigillata is if it soaks in and dries very quickly.
- If there are dark specks in the terra sigillata, put it through a 200-mesh sieve. Even if there aren't dark specks it won't hurt to screen it anyway. It should be about the right consistency to use right away (a specific gravity of 1.15) but if you want it thicker, let it evaporate for a day or two.
- Apply two or three coats to bone-dry ware, allow to dry just until the surface isn't wet, and burnish with fingers, a soft cloth, nylon stocking, chamois-leather, car-polishing mitt or a plastic bag. Don't touch the wet surface and make sure your polishing cloth is very clean. For the greatest shine, apply three thin coats, polishing between coats.
- For colours, add oxides or stains after settling. If you add too much, the terra sigillata won't burnish as well because these materials have a larger particle size than terra sigillata.

Charles and Linda Riggs' recipe

28 lb (16 litres/3½ gallons) water
6.8 kg (15 lb) XX Saggar Clay or OM 4 Ball Clay
1½ tablespoons Sodium Silicate
1½ tablespoons Soda Ash

- Place the water in a 24-litre (5-gallon) pail and add soda ash and sodium silicate.
- Stir in clay – if you have a hydrometer, only add enough clay to get a specific gravity of 1.15.
- Let it settle for 10 minutes then pour off the liquid into a large clear container, discarding the sludge on the bottom of the original container.
- Let it settle for another 20 hours, then siphon the thin liquid from above the layer of sludge. As soon as you notice the liquid getting a little thicker, stop siphoning and discard the rest of the liquid along with the sludge.

Jan Lee's recipes

Ball clay terra sigillata

28 lb (/16 litres/3½ gallons) water
6.35 kg (14 lb) Ball Clay
3.5 tablespoons Sodium Silicate

- Let settle 48 hours before decanting.

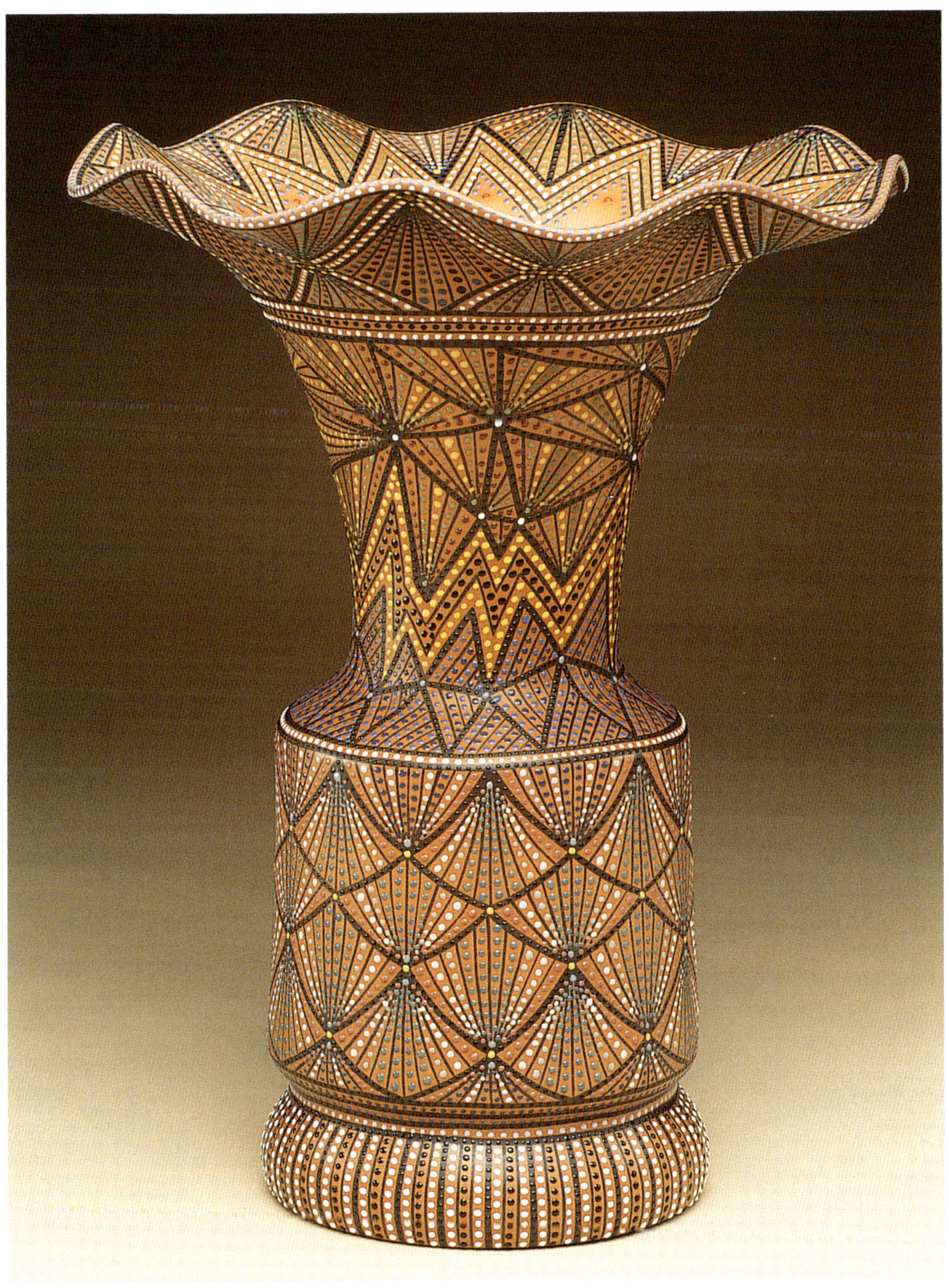

Struck by Ricky Maldonado. Ht: 38 cm (15 in.), dia: 30 cm (12 in.). 'Low-fire terracotta clay, coil-built, sanded using three grades of sanding sponges, coated with six layers of terra sigillata then burnished with hands using the oil from my body (I rub behind my neck or legs). Finished with dry-cleaner plastic wrapped around my fingers. Design drawn in black slip, then dots applied using low-fire glaze between the lines. Fired to cone 06.' *Photo by ImageInation.*

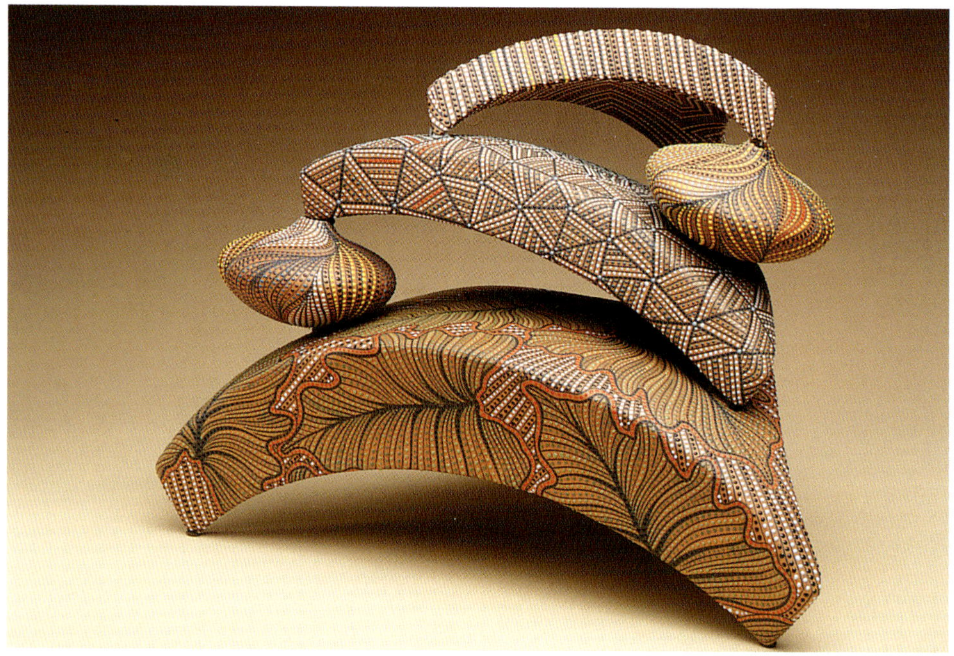

Tri-Becca by Ricky Maldonado. 40 x 43 x 46 cm (16 x 17 x 18 in.). 'Low-fire terracotta clay, hand-built, sanded using three grades of sanding sponges, coated with six layers of terra sigillata then burnished with hands using the oil from my body (I rub behind my neck or legs). Finished with dry-cleaner plastic wrapped around my fingers. Design drawn in black slip, then dots applied using low-fire glaze between the lines. Fired to cone 06.' *Photo by ImageInation.*

Orange terra sigillata

14 lb (6.35 kg) Dry Clay (Orangestone from Highwater Clays)
28 lb (16 l/3½ gallons) water
3 tablespoons Sodium Silicate

- Let settle 6-7 hours before decanting. Allow to evaporate to a specific gravity of 1.15. Jan Lee also sometimes adds crocus martis (a form of iron oxide) to white terra sigillata to make a purplish colour. She uses ½ teaspoon per cup of terra sigillata.

Some of these recipes suggest checking specific gravity with a hydrometer. You can also pour 100 ml (3½ fl oz) of terra sigillata into a graduated cylinder and weigh it – if it weighs 115 grams it has a specific gravity of 1.15. This is the recommended consistency for terra sigillata. If it is much thinner it is hard to apply enough without over-saturating your pot with water. If it is much thicker, it may peel off after firing. In practice you will get to know just how you like your terra sigillata and you won't have to check the specific gravity of every batch you make.

For more information about making terra sigillata, especially if you want to try clays other than those called for in these recipes, I would highly recommend consulting a comprehensive book from the European Ceramic Work Centre called *The Ceramic Process: A Manual and Source of Inspiration for Ceramic Art and Design.*

Troubleshooting

Terra sigillata occasionally suffers from problems adhering to the clay it is applied to. Terra sigillata is not integral to the material the pot is made from, as is the surface of a stone-burnished pot; at the same time it doesn't adhere by melting like a glaze does. Some potters prefer to burnish with a stone for this reason, but you can keep this problem under control by attention to some rules of thumb in applying terra sigillata and firing your ware. Most important is to make sure the terra sigillata is quite thin and watery. It should take several coats to cover your pot, and the final depth of applied terra sigillata should be a fraction of a millimeter. The pot must be dry but not dusty. If necessary, use a damp sponge to remove dust from the surface before applying terra sigillata. However, beware of sponging the surface of your pot too much: every time you rub it with a wet sponge you are removing particles of plastic clay and leaving behind the heavier particles of clay and coarse grog, which don't hold on to terra sigillata as well. After coating your pot, allow it to dry completely before firing; it is a good idea to wait 24 hours.

Avoid a too-rapid increase in temperature during any firing – a pot which comes out of a bisque firing perfectly smooth may peel in a hot and fast pit-firing. If you experience a great deal of peeling despite following all these rules, you may have to change the clay you use or the terra sigillata recipe. One last thing to try is adding a little CMC gum to your terra sigillata. This is a gum which is often used as a binder. It comes as a powder which has to be dissolved in hot water before it can be added to your terra sigillata. Add only a very small amount at a time and test as you go. CMC may reduce the level of sheen you get from your terra sigillata, and it will slow the drying time.

Many potters use terra sigillata as a canvas for the fire to paint on, either pit-firing or saggar-firing their burnished pots. However, other potters like the satiny surface of terra sigillata as is, without any special firing. Ricky Maldonado is a potter

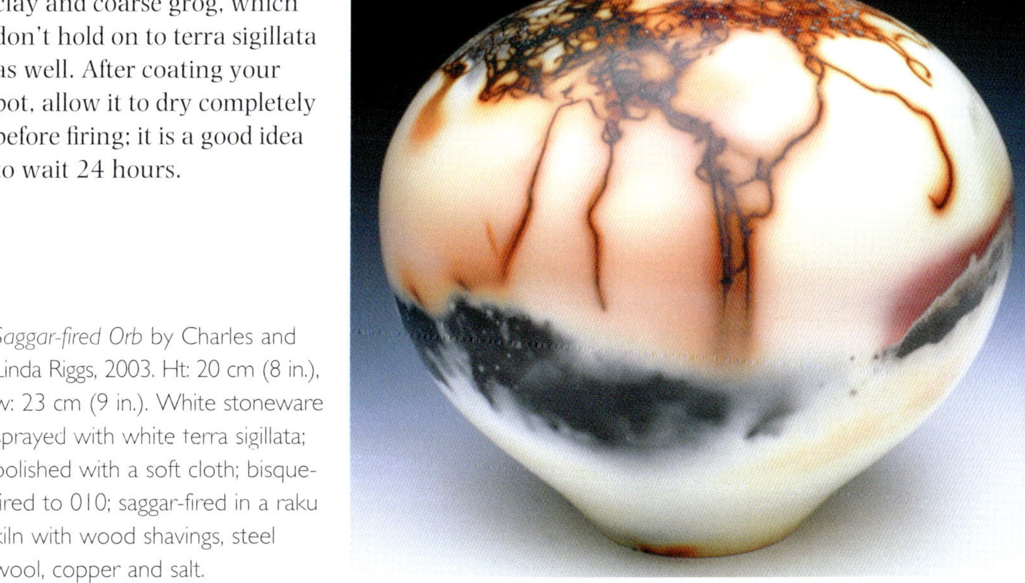

Saggar-fired Orb by Charles and Linda Riggs, 2003. Ht: 20 cm (8 in.), w: 23 cm (9 in.). White stoneware sprayed with white terra sigillata; polished with a soft cloth; bisque-fired to 010; saggar-fired in a raku kiln with wood shavings, steel wool, copper and salt.

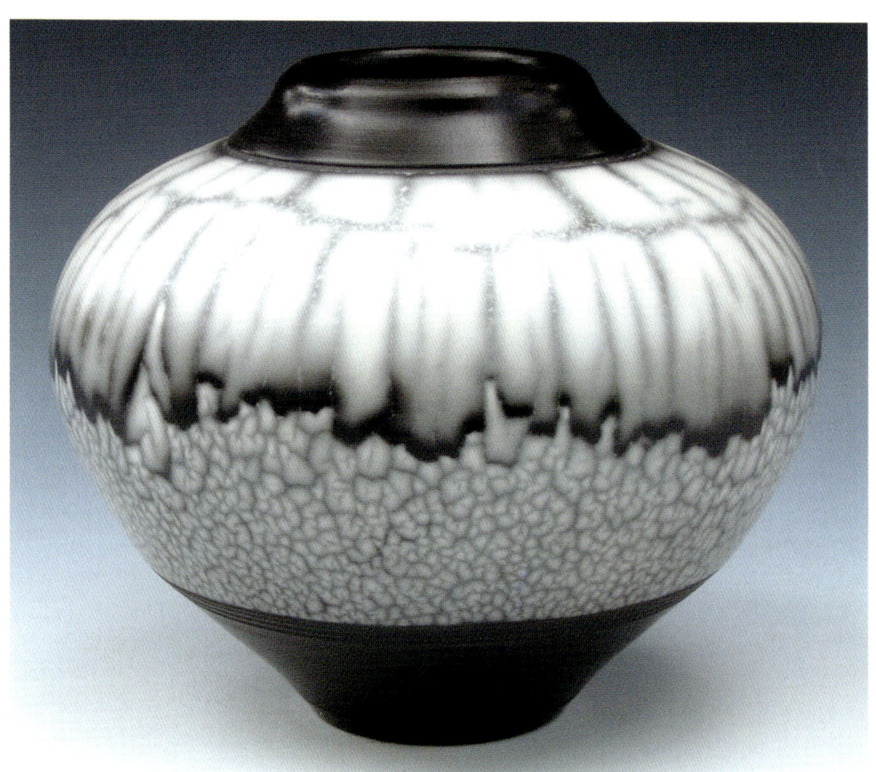

who uses terra sigillata as a base for painting intricate patterns, rendering them with tiny dots of low-fire glaze covering the burnished surface of his pots.

You can also try using terra sigillata to partially cover a surface which has been burnished with a stone. In this way you can have a multi-coloured burnished pot with a high level of sheen. I frequently use terra sigillata made from Cedar Heights Redart clay to paint intricate patterns of fine lines on burnished pots made from red clay. I then smoke-fire these pots to achieve a black-on-black effect. I adopted this process because I found that plain clay slip often rubs off a burnished surface, while the process of buffing terra sigilatta to shine it up seats it even on an already burnished surface.

ABOVE *Naked Raku Orb* by Charles and Linda Riggs, 2003. Ht: 15 cm (6 in.), w: 18 cm (7 in.). Stoneware painted with white terra sigillata and polished with soft cloth; bisque-fired to 010; covered in resist slip and glaze. Sgraffito through glaze before raku firing to 760°C (1400°F). *Photo courtesy of the artists.*

RIGHT, TOP *Spirit Keeper* by Judith Motzkin, 2005. Ht: 35 cm (14 in.). White earthenware, burnished with a rib; terra sigillata brushed on using a banding wheel; bisque-fired to cone 09 and saggar-fired in a gas kiln. *Photo by Judith Motzkin.*

RIGHT *Pele* by Sumi von Dassow, 2008. Dia: 25 cm (10 in.), ht: 18 cm (7 in.). Wheel-thrown B-Mix clay, terra sigillata brushed on and burnished with a chamois-leather; bisque-fired to cone 010; pit-fired with salt-water soaked corn husks. *Photo by Sumi von Dassow.*

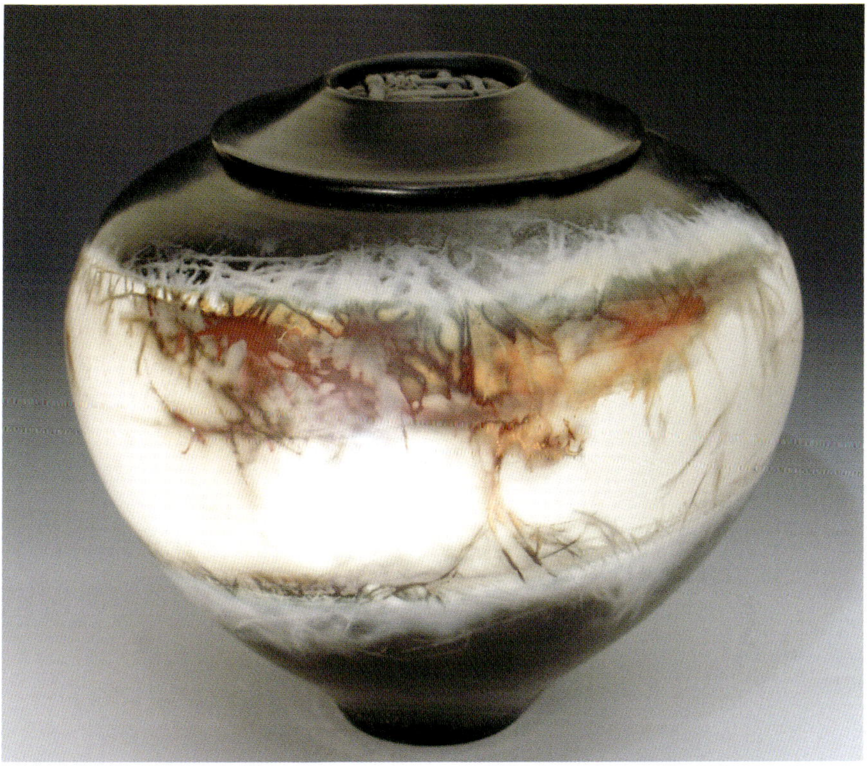

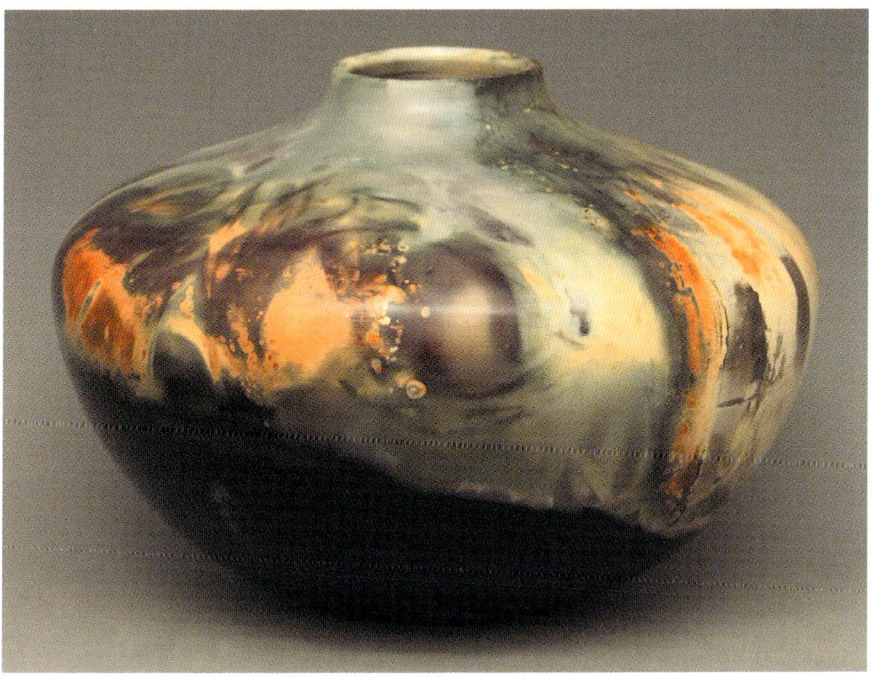

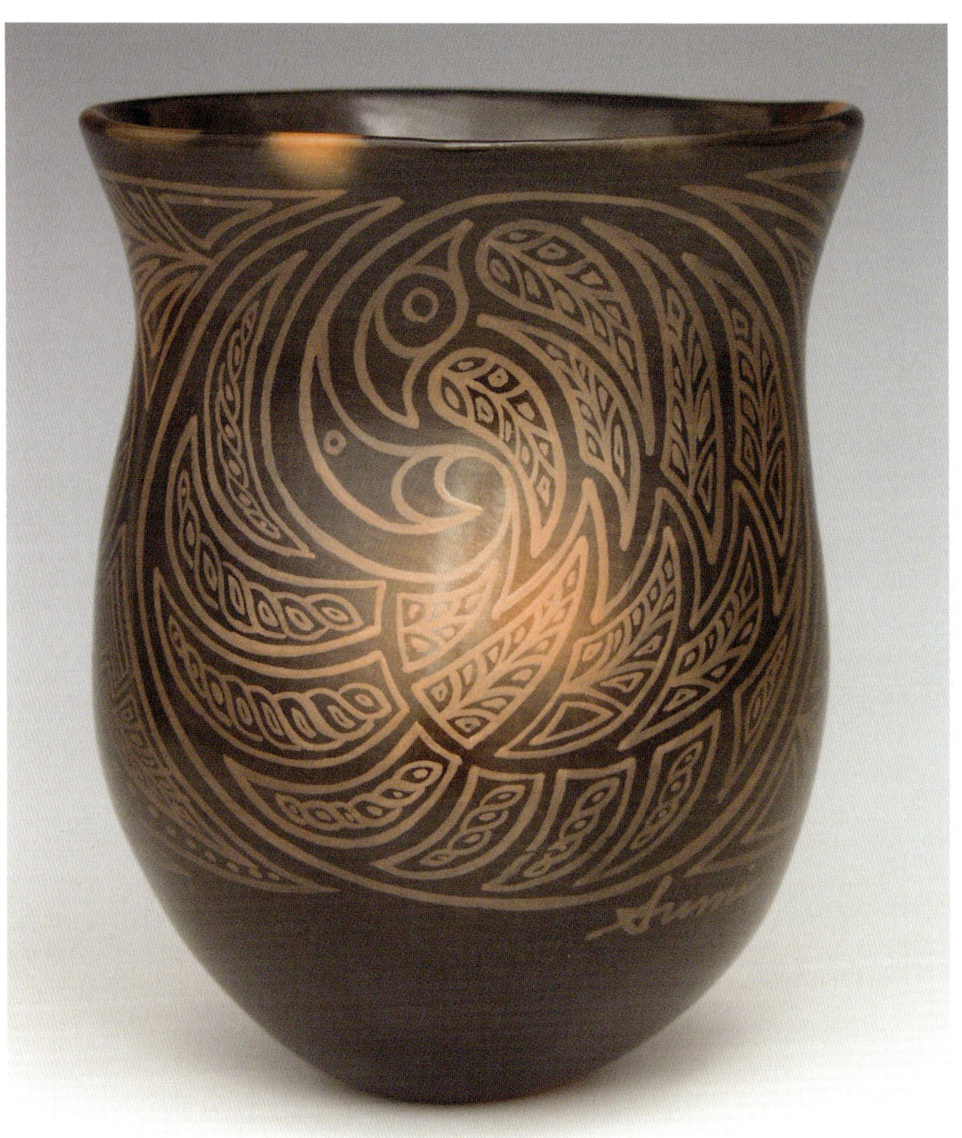

Eagle Vase by Sumi von Dassow, 2008. Ht: 23 cm (9 in.), dia: 18 cm (7 in.). Coil-built from red stoneware (Navajo Wheel Clay from Industrial Minerals Co.); burnished with a stone; decoration painted on using terra sigillata, and smoke-fired. *Photo by Sumi von Dassow.*

RIGHT, TOP *Bowl* by Duncan Ross, 2007. Ht: 17 cm (6$\frac{1}{2}$ in.). Burnished terra sigillata with resist decoration. Smoke-fired several times to different temperatures in a saggar with sawdust.

RIGHT *Bowl* by Duncan Ross, 2007. Ht:16 cm (6$\frac{1}{4}$ in.). Burnished terra sigillata with resist decoration. Smoke-fired several times to different temperatures in a saggar with sawdust. *Photos by Duncan Ross.*

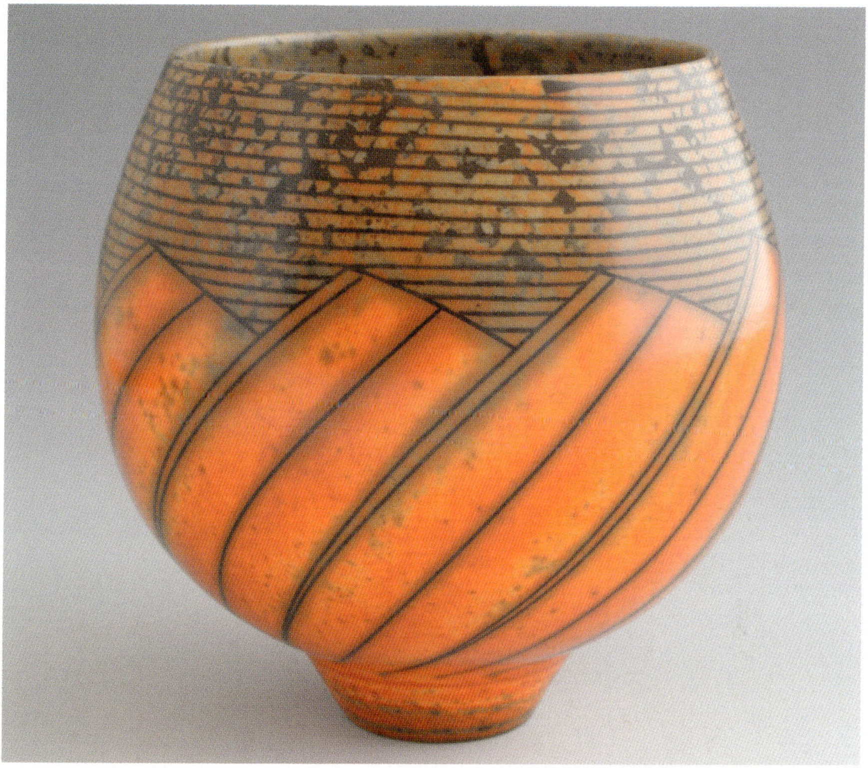

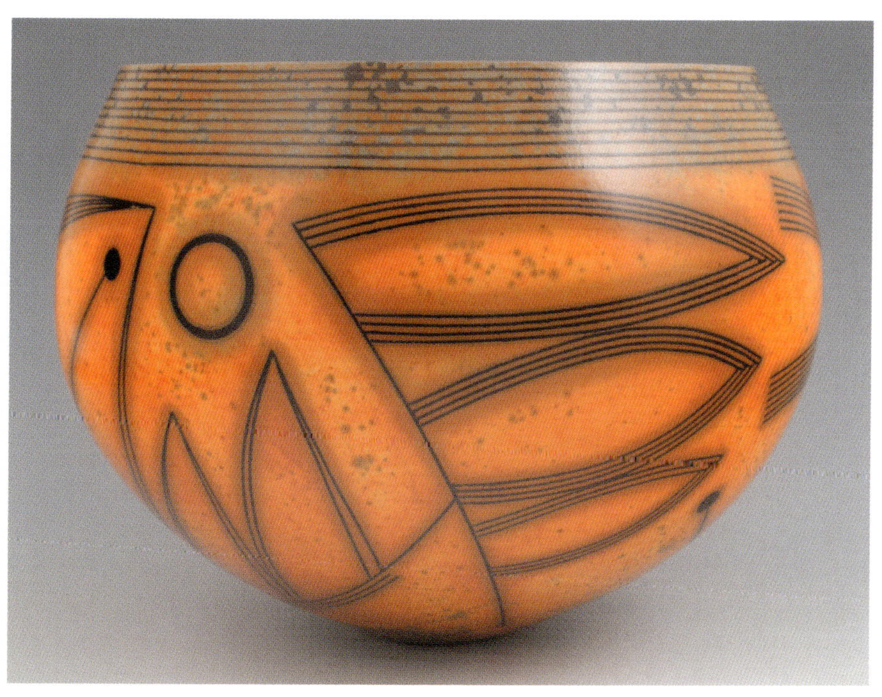

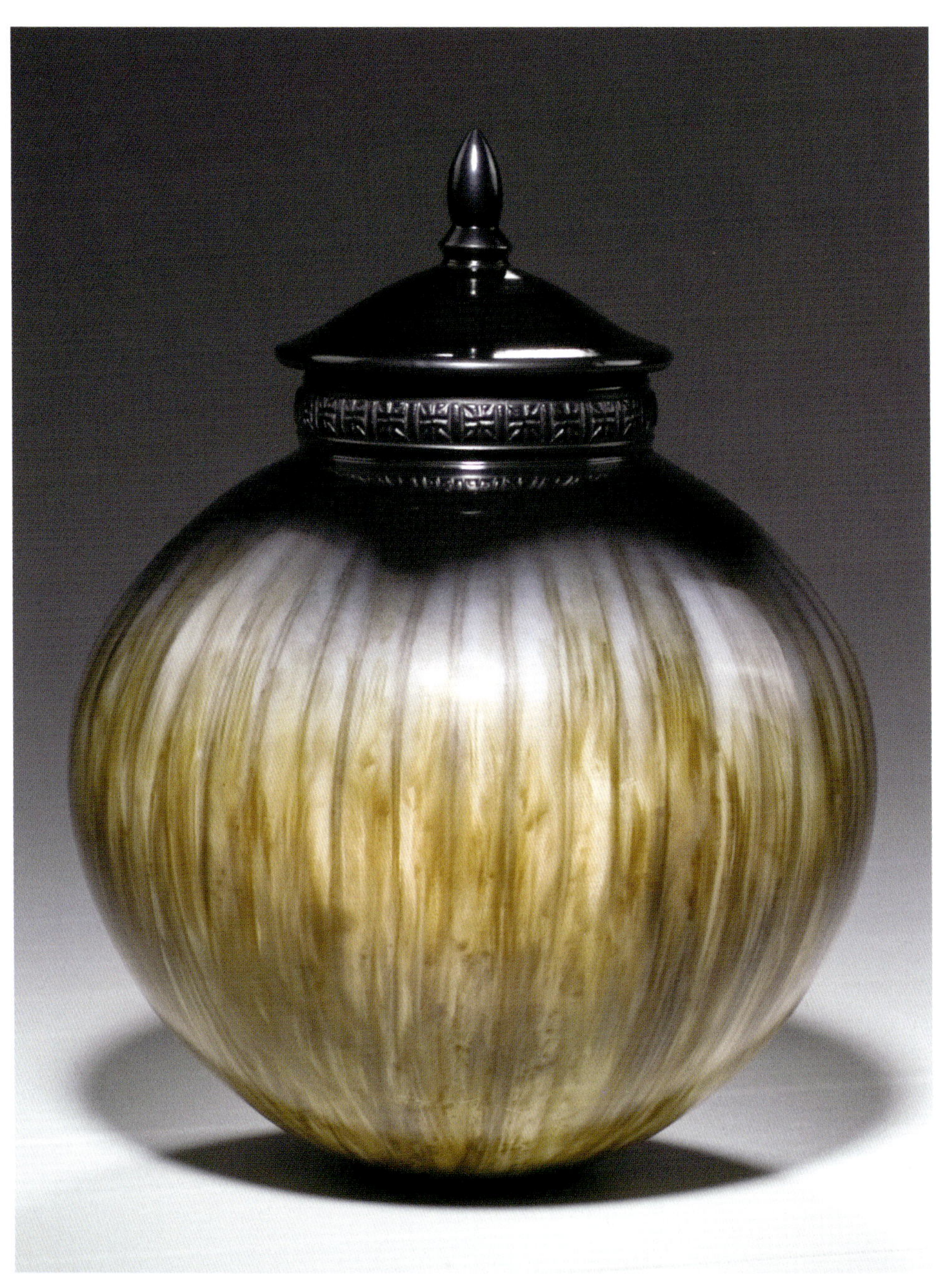

Covered Jar by David Greenbaum Ht: 33 cm (13 in.). Black-fired in an Axner Super Kiln in a saggar, partially buried in sawdust; then pit-fired wrapped with broom corn and grasses. *Photo by Randy Batista.*

Chapter 4

Smoke-firing and black-firing

One of the most striking ways to finish a burnished pot is simply to fire it black. This technique is most famously used by Native American potters of the southwest USA, though in various other parts of the world traditional potters have also used smoke to blacken burnished pots. Essentially, to turn pots black all the potter needs to do is to smother the fire when it is hot, and prevent oxygen from touching the pots before they have cooled. This process can take place in a kiln, in an oil drum or trash can (metal dustbin) or in an open bonfire; it can be done with pots which have been previously bisque-fired, or with preheated dry burnished greenware.

A very simple way to blacken pots is in what might be called a 'modified saggar' firing. After bisque-firing a burnished pot to cone 018, wrap it in newspaper, then in a double layer of aluminum foil, and fire it in a kiln to cone 021. This is a low enough temperature that the foil doesn't begin to disintegrate, but hot enough to turn a suitable clay a rich black. (Some clays won't pick up enough carbon from burning newspaper alone at such a low temperature; you'll just have to try this technique with your own clay body. You have a good chance of success using a red cone 6 clay body.) This technique can be used in an electric, gas or raku kiln. Remember that it is bad for the elements of an electric kiln to burn anything in it, though in practice an occasional smoke-firing with newspaper won't appreciably harm them, especially if such reduction firings are alternated with normal firings, or the kiln is adequately ventilated.

However, there are more exciting ways to blacken pots – and combustible materials other than newspaper will burn at a higher temperature and may turn your pots a richer black. If you have access to a raku kiln, of course, you can easily use the standard raku process to blacken your pots. However, the surface of a burnished pot is very likely to get damaged if it is picked up with tongs, so this method must be used very carefully. Actually, since you don't need to achieve a temperature hot enough to melt glazes, it is possible to dispense with the raku kiln and simply blacken your pots in a barrel directly. I developed a method of doing so which is reliable, dramatic, quick and won't alarm the neighbours too much if you are firing in an urban area.

I start by placing a layer of newspaper, sawdust and wood scraps on the bottom of a trash can (or metal dustbin). A barbecue grill is elevated over this pile of combustibles using bricks or kiln posts, then three to five pots (bisqued to cone 018) are placed on the grill. Since the firing is quite fast, the pots are preheated in an electric kiln to about 120°C (250° F) before placing them in the trash can. To protect the pots, each one is covered with a coffee tin or a large popcorn or biscuit tin, with lots of holes punched in it for the smoke. Larger pots may be fired one at a time, and if a large enough tin can't be found the pot can be protected inside a cage of hardware cloth or wire mesh (sturdier than chicken wire). I build a fire around and under the tins with newspaper and kindling, adding larger pieces of wood as the fire takes hold. At this point, to raise the temperature of the fire, I literally fan the flames by holding an electric fan over the trash can. Once the fire is burning merrily, I throw in a handful of damp wood shavings wrapped in a sheet of newspaper, then cover the trash can with an airtight lid. After cooling for an hour or so, it can be opened.

It is even possible to use sawdust to blacken pots inside an electric kiln without damaging it – if you use a kiln specially designed for the purpose, as David Greenbaum does. His Axner Super Kiln is manufactured with sturdy elements and is coated on the inside with ITC 100 coating to protect the elements from damage due to smoke and reduction.

He puts the pot to be blackened on a kiln shelf inside a custom-made stainless steel saggar with sawdust underneath the pot but not touching it. This saggar is placed inside the kiln, covered with a lid, and fired slowly to about 760°C (1400°F). As a variation, if he doesn't want the entire pot black, he will partially bury it in vermiculite inside the saggar to keep that portion of the pot white. These partially blackened pots are finished in a subsequent firing by wrapping them in grasses or pine needles and setting this bundle on fire. The burning materials leave their imprint on the unblackened areas without affecting the blackened portion of the pot.

Michael Wisner fires his burnished pots made from white earthenware using a method he adapted from Native American firing techniques. He places warmed bisque-fired pots on stands over a bed of sawdust, then covers them with a metal bucket or oil drum, carefully sealing around the base of the bucket

1 The metal bin or trash can is prepared by placing a barbecue grate on kiln posts, with a small pile of wood shavings, paper and kindling underneath. The preheated pots are placed in the bin on top of the grate, and each pot is covered with a coffee tin pierced with holes.

2 The paper and kindling around the pots are lit.

3 Blowing air into the metal bin/trash can with a table fan encourages the flames and raises the temperature.

4 When the flames can be seen around and under all the pots and a good flame is going, it is nearly time to cover the bin/trash can.

5 A packet of damp shavings is added to the flames as the last step before covering the bin/trash can.

6 As soon as the packet of shavings starts to burn the lid is put on to smother the fire and create lots of smoke around the pots.

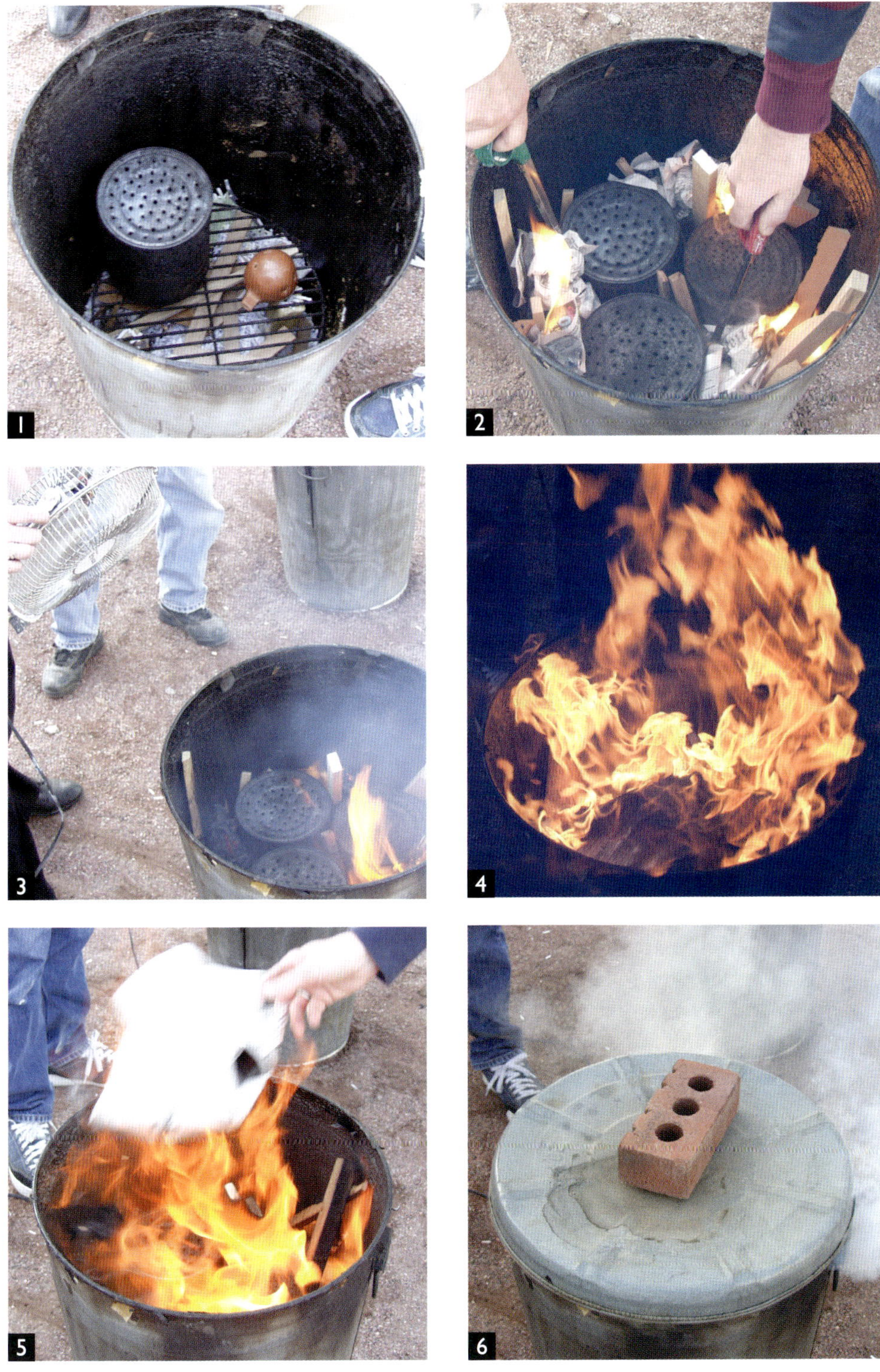
1
2
3
4
5
6

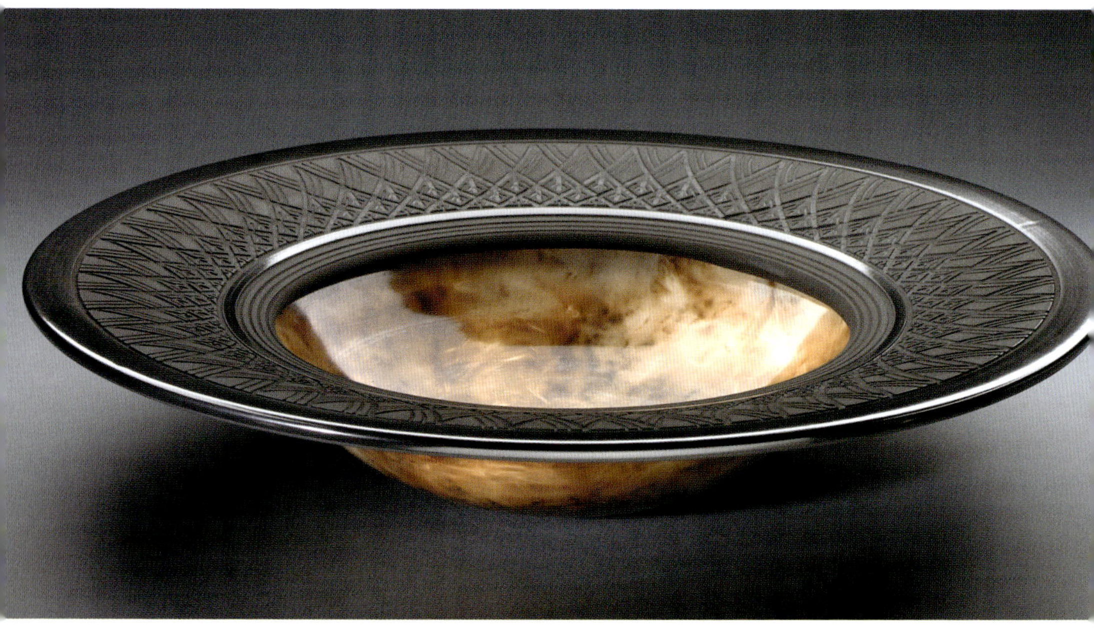

Large Bowl by David Greenbaum. Dia: 59.5 cm (23.5 in.). Black-fired in an Axner Super Kiln in a saggar, partially buried in vermiculite; then pit-fired wrapped with broom corn (straw purchased from a craft supplier for making brooms) and grasses. *Photo by Randy Batista.*

with sand or sawdust. He stacks firewood around the bucket, holding it in place with baling wire, lights it, and allows it to burn for 20 minutes or so. The fire burning around the bucket creates enough heat to cause the sawdust inside to smoulder, blackening the pots. This fire doesn't need an extra air source to get hot enough, since the fire around the bucket is not enclosed, but you do need to choose your firing site carefully.

A gas kiln is a handy piece of equipment for blackening pots using a saggar or a barrel. To fire inside a kiln, Michael Wisner uses a specially designed firing rack to stack pots over a bed of sawdust or manure, covers them with a trash can, and fires to cone 012. You could also build a box out of soft bricks inside your kiln and fill it with sawdust and pots, or make saggars out of coarse clay to fit one or more pots. Your combustible material can be sawdust, wood shavings or even manure. As long as your saggar is airtight, the pots inside will turn black. Ian Garrett blackens pots in individual saggars, burying them in sawdust and taking seven hours to achieve a final temperature of about cone 010. He uses such a prolonged firing because he doesn't bisque-fire, and to achieve a richer black.

Wisner learned the techniques he uses for burnishing and firing from Juan Quezada of Mata Ortiz. Quezada mastered traditional techniques to create both black and polychrome burnished pottery, then taught these techniques to other potters in the town. As a result, Mata Ortiz is now internationally famous for its burnished

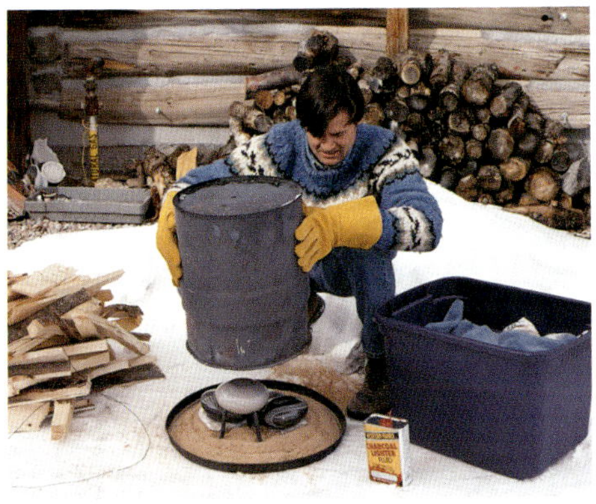

Michael Wisner preparing to black-fire burnished pots in an oil barrel: oven-warmed bisque-fired pots placed on firing stands over a bed of sawdust are covered by a barrel.

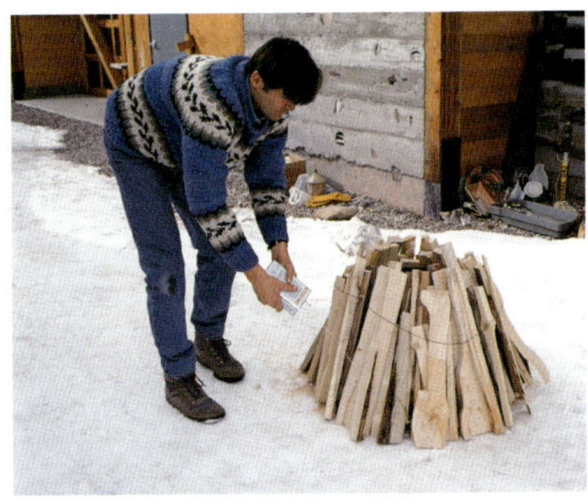

The barrel is surrounded by wood tied in place with baling wire.

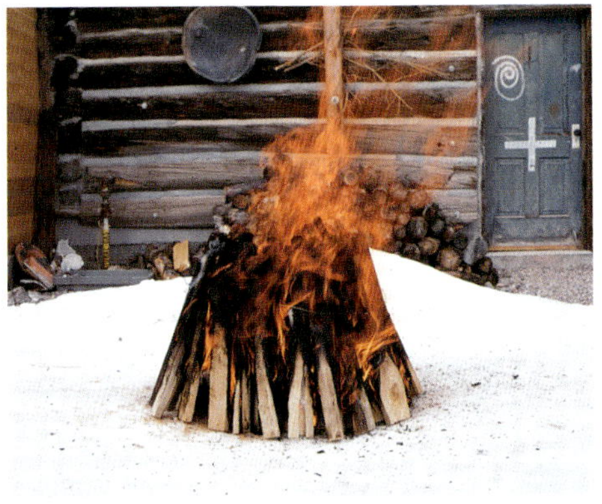

The wood burns freely, causing the sawdust inside the barrel to smoke.

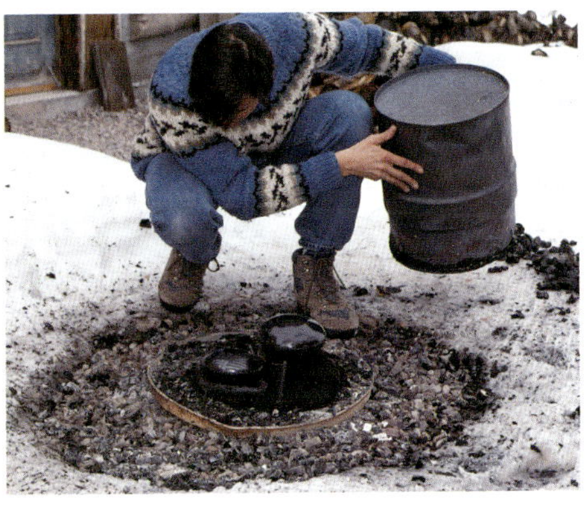

After the fire burns down, Michael rakes away the coals and lifts the barrel to reveal shiny black pots.

pottery. Potters use locally available materials to fire their pots on the open ground. While each potter adapts the materials and equipment he can find to his own uses, a typical method of firing starts with perching an unfired pot on three stones over a bed of ground-up manure. Other potters may place one or more pots on a metal rack – generally something salvaged or improvised, such as a bicycle wheel. Whether propped on stones or on a bicycle wheel, the pots are above, but never in direct contact with, the manure. A large cover pot is then placed upside-down over the unfired pot, then surrounded by dried cowpats or firewood. The fuel is lit, and after it has burned down the ashes are scraped up to completely seal the gap around the bottom of the cover pot or bucket. In this way the pot never touches the burning fuel, but the manure under the pot produces enough smoke to blacken it. The ashes insulate the pot and keep it from re-oxidizing as it cools. Because the pots are not bisque-fired before this firing they emerge a rich, brilliantly shiny black.

Mata Ortiz potters are also famous for polychrome pottery, in which coloured slips are painted over the burnished surface. To fire polychrome pots the potters use a similar technique, but leave a hole in the cover pot to allow oxygen to circulate around the pot during the firing, and they don't insulate the pot

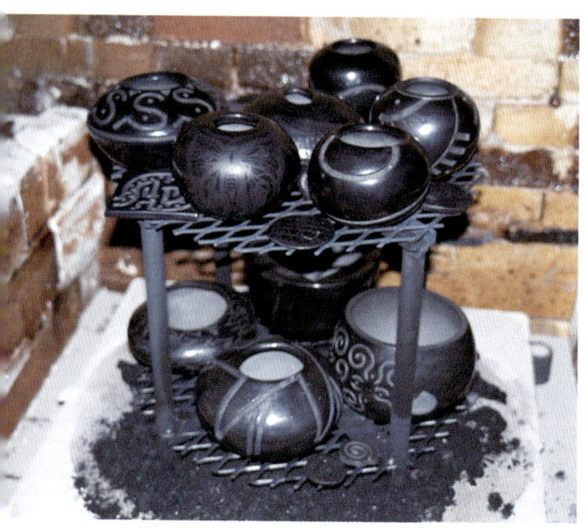

Black pots ready to unload from Michael Wisner's gas kiln. They were placed on a steel firing rack over a bed of manure, covered with a steel drum, and fired to cone 012.

BELOW *Black Diamond Donut* by Michael Wisner, 2008. Dia: 18 cm (7 in.), ht: 10 cm (4 in.). Burnished using tights instead of a stone and smoke-fired.

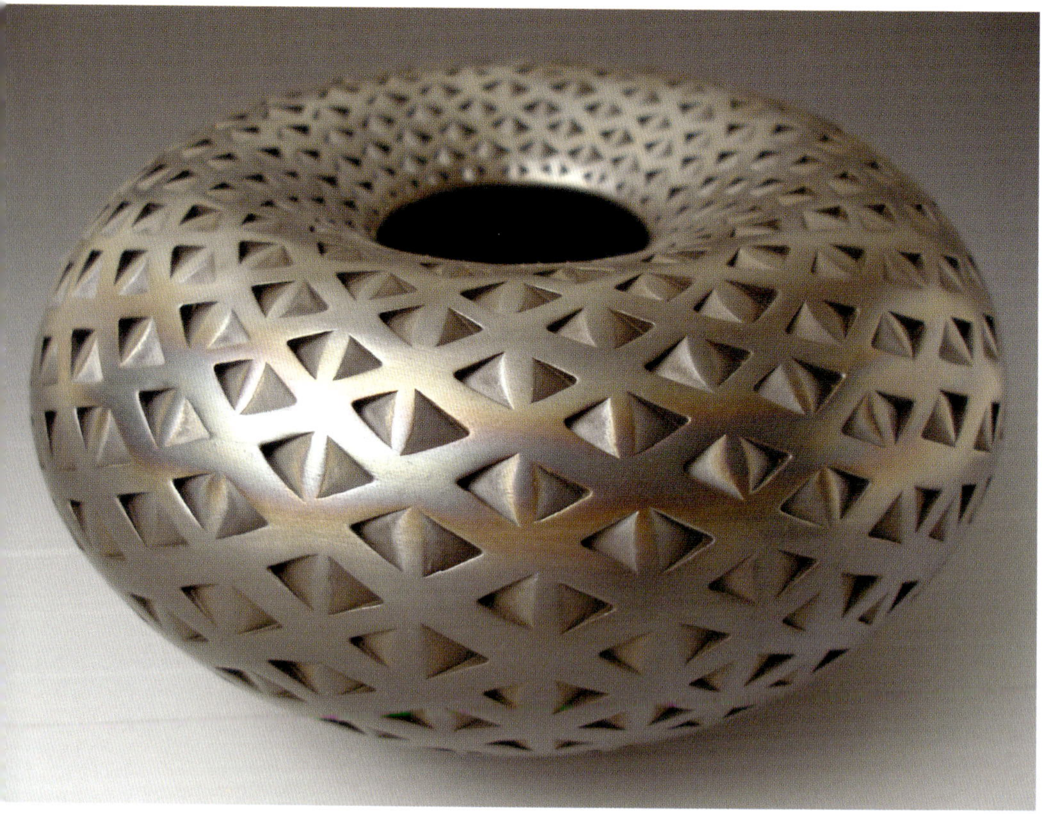

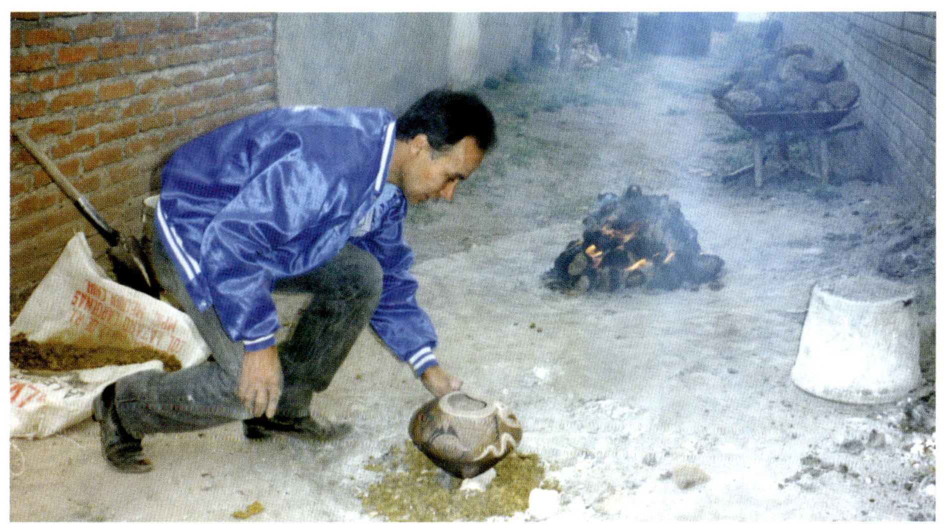

ABOVE Rito Talavera prepares to fire pots for Lydia Quezada in Mata Ortiz. He carefully perches a pot on three rocks over a pile of pulverized manure. It will be covered with a larger pot, surrounded by cow manure and lit. *Photo by Bill Hensler.*

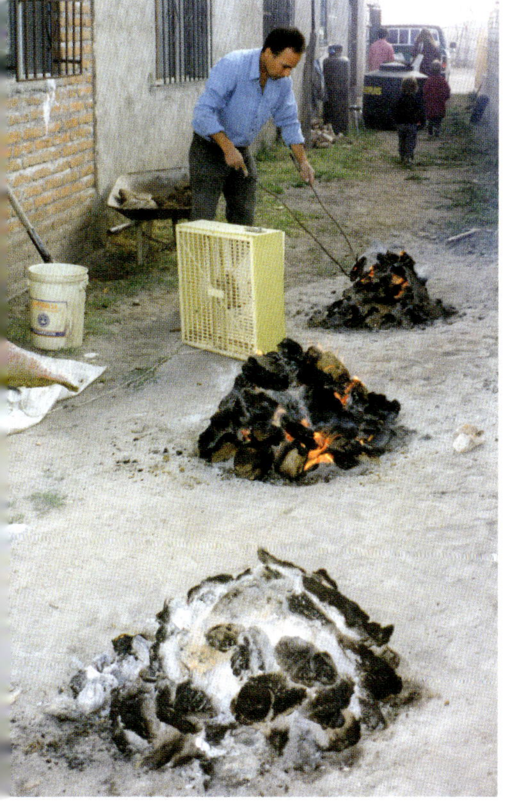

LEFT Each mound of burning cow manure contains one pot being blackened. Notice the use of a fan to raise the temperature. On a windy day the fan might not be needed. *Photo by Bill Hensler.*

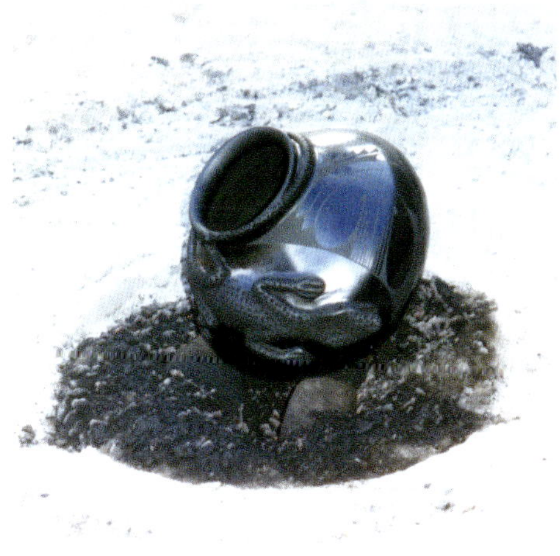

ABOVE The finished black pot, uncovered, still perched on three stones. *Photo by Bill Hensler.*

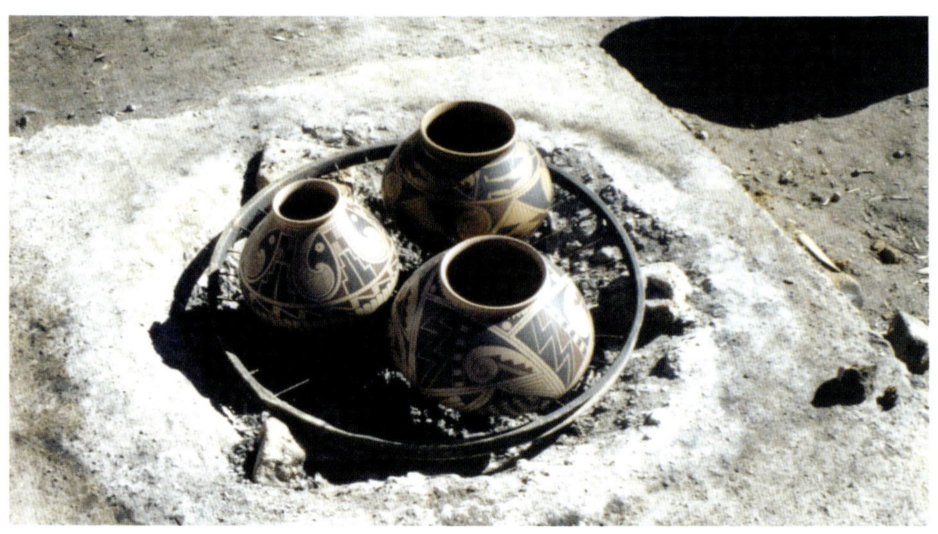

Polychrome pots by Sabino Villalba of Mata Ortiz, cooling after firing. These pots were fired on a bicycle wheel propped on rocks, under a bucket with a hole in it so that they wouldn't become black. *Photo by Bill Hensler.*

with ashes after the fire dies down. These traditional methods of firing are very risky compared to firing in a kiln, and many pots do not survive the intense fast heat rise – only those pots whose walls are perfectly even and thin survive, and they are highly sought after by collectors.

Pueblo Indians use similar techniques to build, burnish and fire their pots. A renaissance of the Pueblo pottery tradition took place in the first half of the 20th century, sparked by several women potters who developed their skills and techniques and passed them on to their descendants. Nancy Youngblood, of Santa Clara Pueblo, is the granddaughter of Margaret Tafoya, one of these founding grandmothers of Pueblo pottery. Nancy is one of the most famous Pueblo Indian potters working today. Her highly sought-after pots sell for thousands of dollars each and are often sold before they are even fired. Preserving the Pueblo Indian tradition, she fires her pots in a bonfire, which can be risky but results in gloriously glossy black or red pots. In order to ensure success in each firing, she designed and built a shed where her pots can be safely fired in any weather. She preheats the pots in an oven before placing one pot at a time on a specially designed rack in the centre of a large sand-filled firing area. She first builds a small fire under the pot with wood chips to heat it slowly, then surrounds the pot and rack with planks, allowing the fire to achieve the necessary temperature to mature the clay properly. The pot, positioned on the firing rack, is completely surrounded by the fire and smoke without coming into any contact with the burning fuel. If the pot is to be red, the ashes are raked away from the fire as the pot cools, so that the pot doesn't pick up any smoke clouds. If the pot is to be black she smothers the fire with sand.

TOP TO BOTTOM

One of Nancy Youngblood's pots in the early stages of the firing process, on a firing rack with a small fire under it to heat it slowly. The rack protects the pot from direct contact with burning wood while allowing smoke to freely circulate around the pot. The metal cover on the rack protects the pot when the fire is smothered.

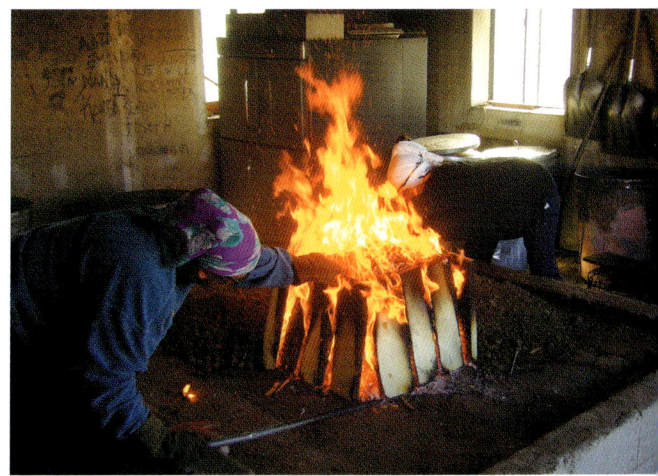

After slowly building up the fire under the pot with small pieces of wood, Youngblood and an assistant surround the firing rack with boards and begin the process of smothering it. The boards protect the pot from contact with the sand used later on, and produce the smoke needed to blacken it.

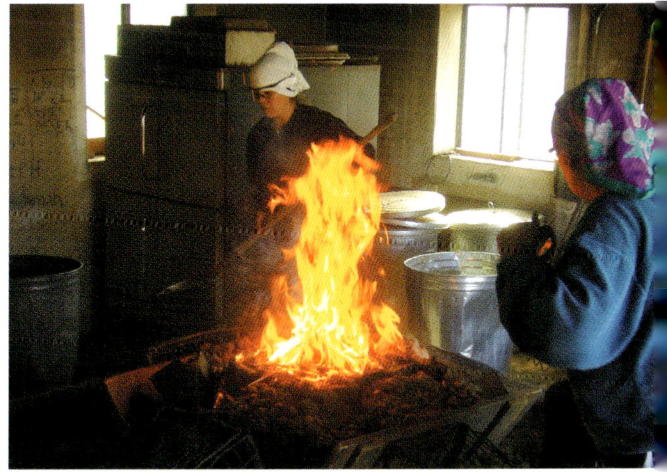

Youngblood and an assistant smother the fire with copious amounts of sand. The firing has taken about two hours at this point. After the mound cools, the sand will be raked away and the blackened pot will be removed from the rack.

Photographs by Greg Howett.

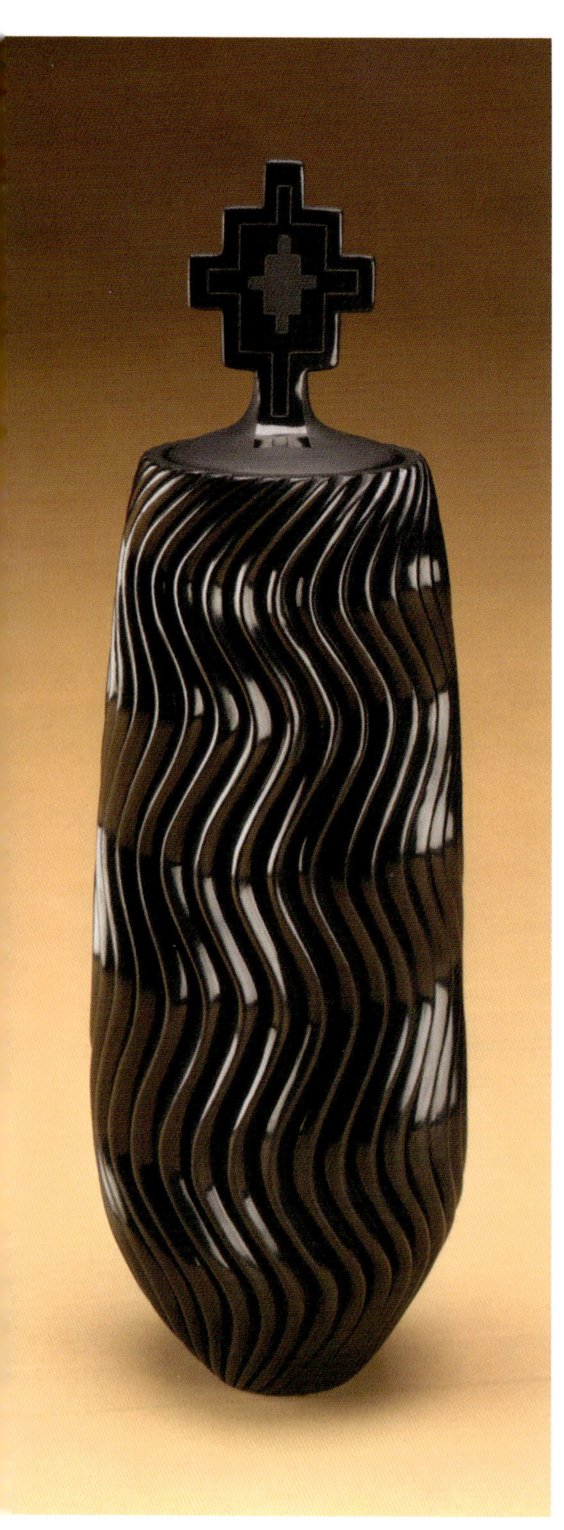

Native American potters who fire in bonfires generally consider any unintentional clouds of smoke to be a defect, but potters from European traditions have enthusiastically adapted open bonfire methods of firing deliberately in order to produce random markings from smoke. Inspired by the search for such random smoke-markings, pit-firing and saggar-firing techniques were born. Pit-firing now flourishes, particularly in the USA, as a popular firing method which not only yields spectacular results but is great fun to do.

LEFT *Tall Swirl Vase with Kiva Step Lid* by Nancy Youngblood of Santa Clara Pueblo, 1997. Ht: 35 cm (14 in.), w: 13 cm (5 in.). Coil-built from clay collected on Pueblo land, carved, coated with red slip, burnished with a stone and smoke-fired. *Photo by Hawthorne Studio, Santa Fe NM.*

RIGHT, TOP *Strange Fruit* by Ian Garrett, 2008. Ht: 25 cm (10 in.). Saggar-fired.

RIGHT Round carved form by Ashraf Hanna, Ht: 34 cm (13½ in.). Handbuilt, partially burnished, blackened in post-firing reduction.

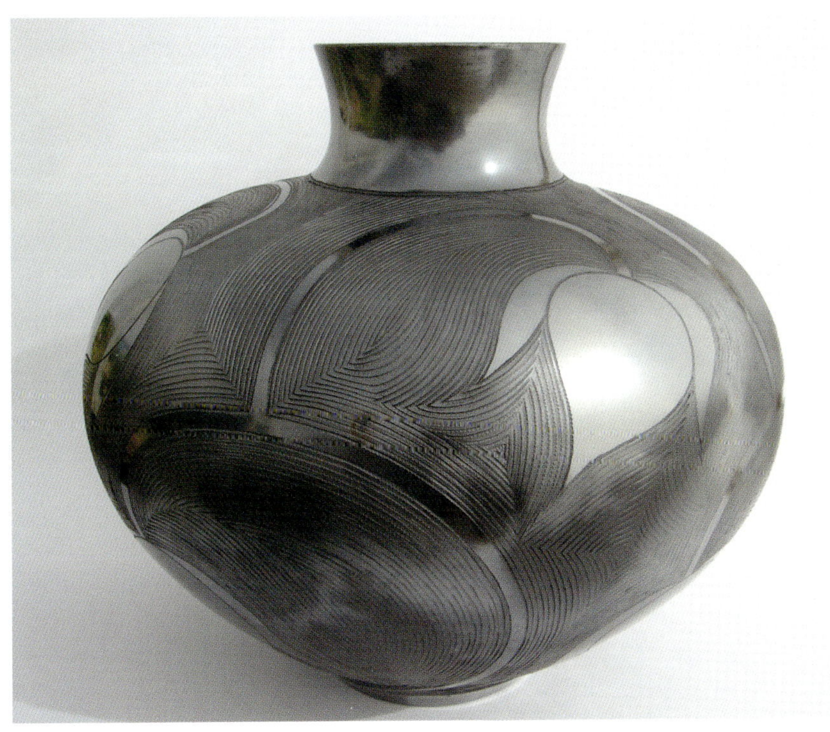

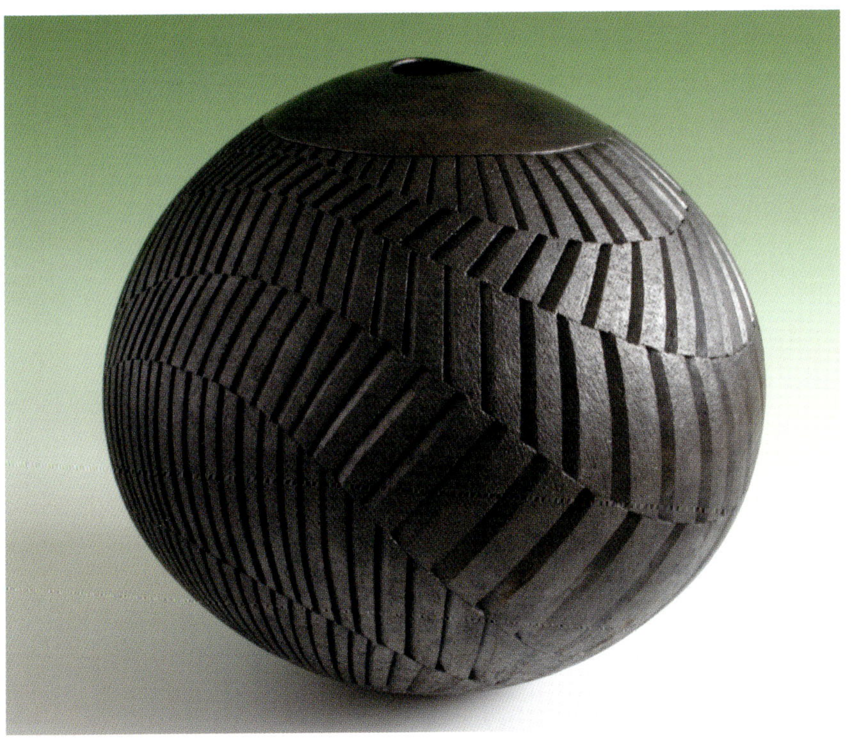

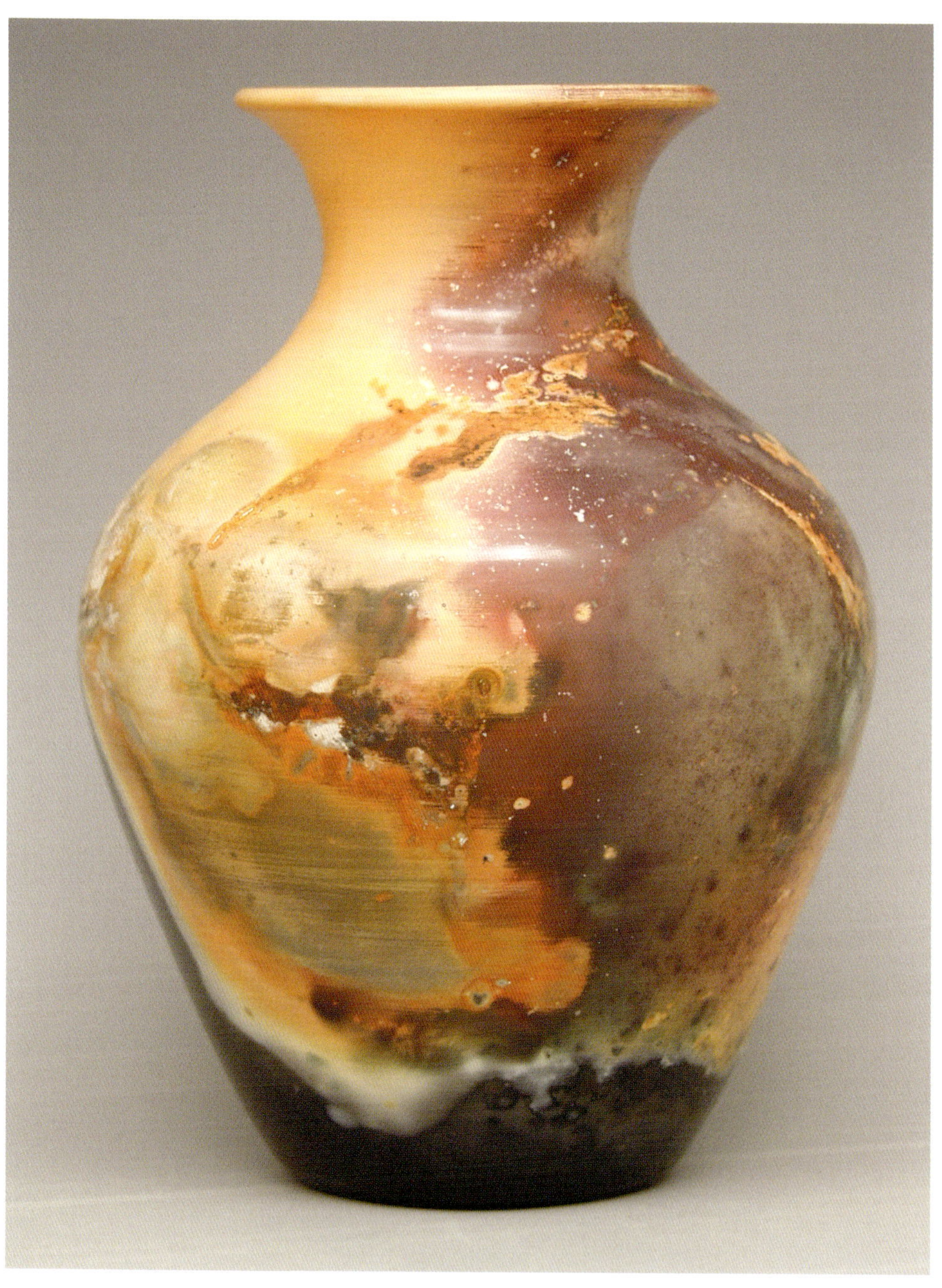

Pit-fired Vase by Sumi von Dassow, 2007. Ht: 23 cm (9 in.), dia: 18 cm (7 in.). Wheel-thrown from B-Mix clay (white stoneware), terra sigillata brushed on and burnished with a chamois-leather; bisque-fired to cone 010; pit-fired with salt, copper carbonate, coffee grounds and seaweed. The coffee grounds and salt make the wash of orange-yellow, the copper carbonate makes the burgundy patches, and the seaweed makes the distinct orange dots. *Photo by Sumi von Dassow.*

Chapter 5
Pit-firing

Pit-firing is one of the most exciting ways to fire a burnished pot. In a pit-firing, pots pick up black and grey marks from smoke and contact with combustible materials, but they can also pick up vibrant warm red, yellow and pink colours from chemicals added to the pit. Pit-firing isn't any single technique – there are as many ways of pit-firing as there are potters who do it. Every potter ends up developing a personal style of pit-firing, depending on variables including the type of clay used, the fuel available, the chemicals added and the altitude at which the firing takes place. A pit can be small enough for one person to fill it, or large enough for hundreds of pots. Pit-firing generally takes at least a few hours and usually must cool overnight, compared with the relatively quick process of black-firing.

Ian Garrett of South Africa uses a simple method of pit-firing, very similar to some of the smoke-firing methods in the previous chapter, to achieve pots marked with smoke clouds. He fires greenware in a shallow pit about 20 cm (8 in.) deep and wide enough to contain one to three vessels. Firing only a few vessels at a time allows maximum surface contact between the work and fuel, giving him control over the degree of surface colour variation. Vessels are first pre-heated slowly up to 200°C (392°F) in an oven or near a small fire, and left for at least an hour to drive off as much moisture as possible. Meanwhile, the pit is lined with a 10-15 cm (4-6 in.) thick layer of sticks and twigs loosely packed to allow sufficient air to penetrate and aid combustion under the stack.

The pre-heated vessels are inverted onto a sheet of metal or clay that has been raised on bricks. Clay shards are placed around the vessels to protect them from rapid heating and excessive fire-flashing. This stack is covered with a densely packed layer of wood bark, dried aloe leaves, cow dung or soft fibrous wood that smoulders slowly before igniting. These soft fibrous fuels allow for a gradual, even increase in temperature. Newspaper and kindling are packed around the sides of the pit and set alight. Once the layer of soft fuel is burning strongly, any available fuel, including hard wood, is carefully thrown on to maintain the fire and build up a layer of coals to surround the pots. The fire is then allowed to burn out, and often the whole pit is covered with a sheet of corrugated zinc to slow the cooling and protect the pots from breezes.

Most potters who pit-fire add chemicals to create colour, not just smoke markings. The essential ingredients of a successful pit-firing consist of burnished pots, usually bisque-fired to cone 010; wood-shavings; chemicals such as salt, copper carbonate or copper sulphate, ferric chloride or

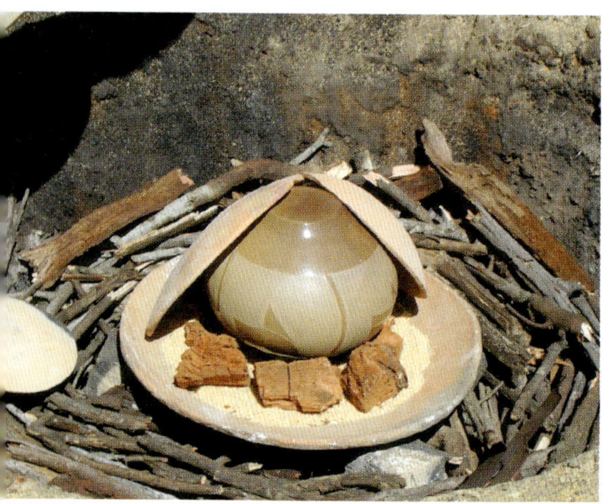

Ian Garrett fires one pot at a time, placing it on a base-shard and protecting it from the direct flame with fired shards.

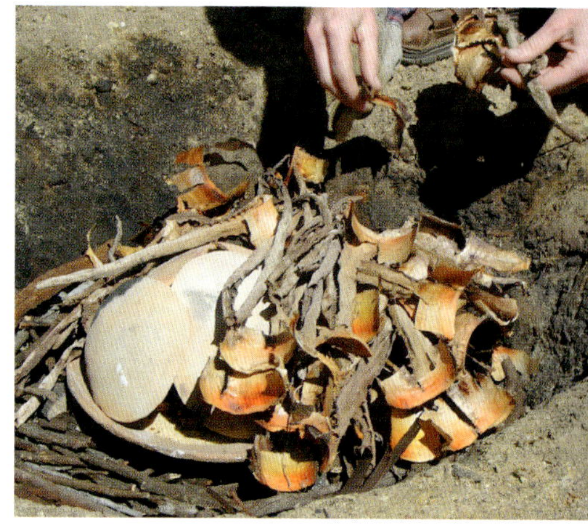

Garrett covers the pot with dried aloe leaves. After he lights the fire he will add rotten wood.

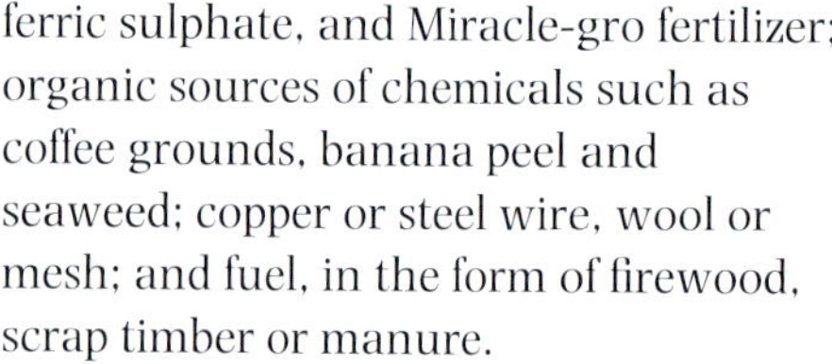

ferric sulphate, and Miracle-gro fertilizer; organic sources of chemicals such as coffee grounds, banana peel and seaweed; copper or steel wire, wool or mesh; and fuel, in the form of firewood, scrap timber or manure.

Usually a firing pit is dug 0.5–1 m (1½–3 ft) into the ground, often into the soft sand of a beach, though a more permanent pit can be built from bricks above ground, or dug partway and heightened with a row or two of bricks, depending on how hard it is to dig deep enough. A rectangular shape is preferable, to create a natural draught, allow access for loading from the sides, and to make it easy to cover the pit during and after the firing. The size can range from 1 m (3 ft) square up to 6 m (20 ft) long or more – a smaller pit is possible, but may not get hot enough to develop the most brilliant colour.

The basic method of pit-firing is to cover the bottom of the pit with a layer of sawdust or wood-shavings, with salt and other chemicals sprinkled on and

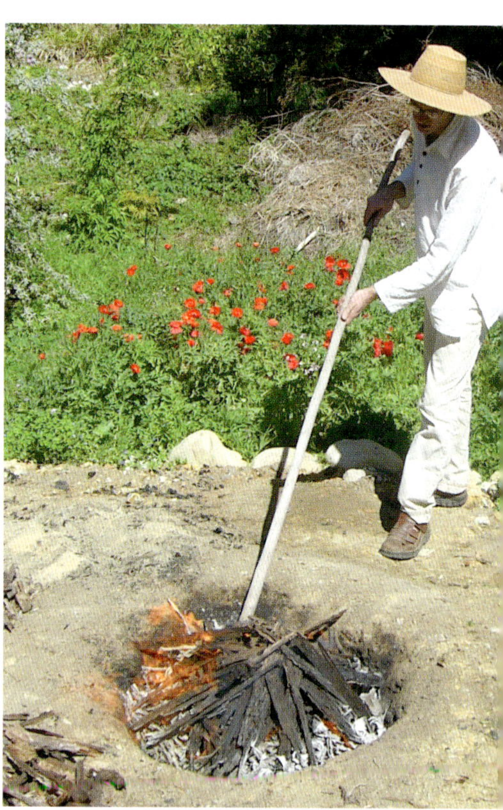

Ian Garrett adjusting the coals to avoid exposing the pot to the cold air.

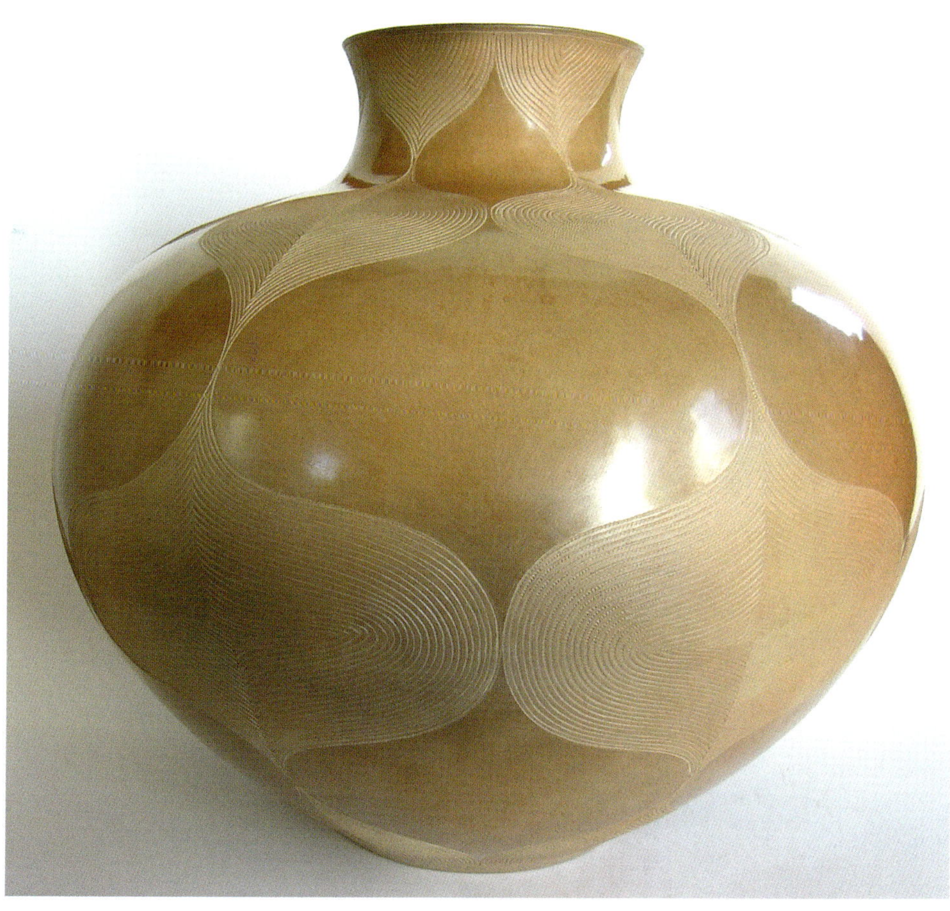

Bulbs by Ian Garrett, 2008. Ht: 28 cm (11 in.). Coil-built from red earthenware, impressed with the serrated edge of white mussel shells when leatherhard, and burnished with oil when dry.

perhaps lightly stirred in. The pots are placed on and in the shavings. If materials such as copper wire or banana peel are used, they may be placed on or tucked between pots, or wrapped around the pots before they are placed in the pit. Pots may even be packed into saggars in the pit, or large pots can be used as saggars for smaller pots. Finally, fuel is piled onto the pots and lit. Frequently, a pit is covered after the fire gets going, to keep the heat in and create some reduction for colour development. The fire may be re-stoked, or it may simply be allowed to burn itself out naturally. In general, a larger and longer pit will develop better colours than a small one. A long rectangular fire will develop a natural draught – especially if you pay attention to prevailing wind patterns – which allows the temperature to reach around 650-760°C (1200-1400°F). This is why potters commonly fire on a beach: not only is it easy to dig a large pit in the sand, but the wind is often reliable and predictable on a beach,

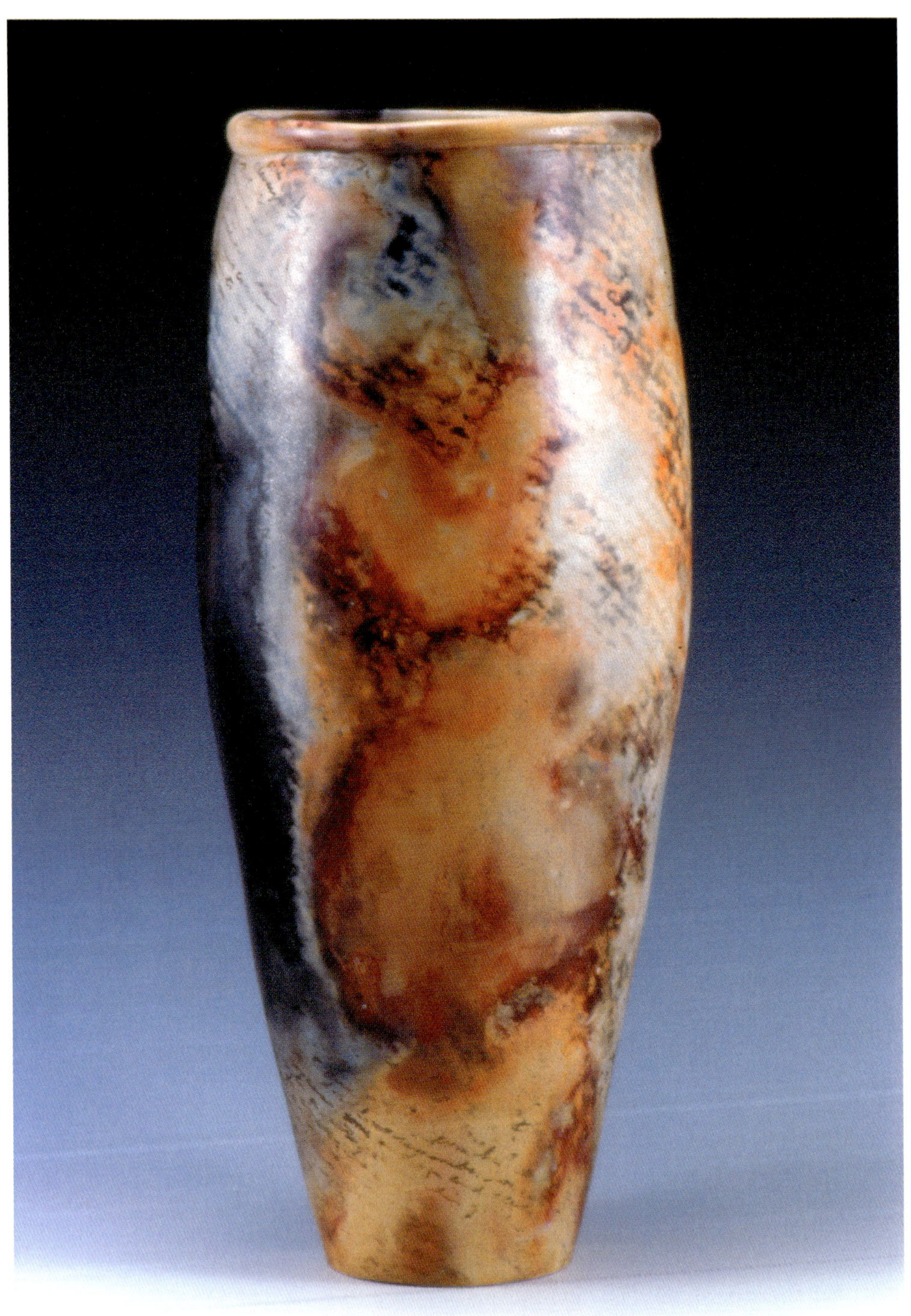

Remaining Voices by Jane Burton. Ht: 62 cm (24½ in.), w: 23 cm (9 in.). Pit-fired vessel with writing.

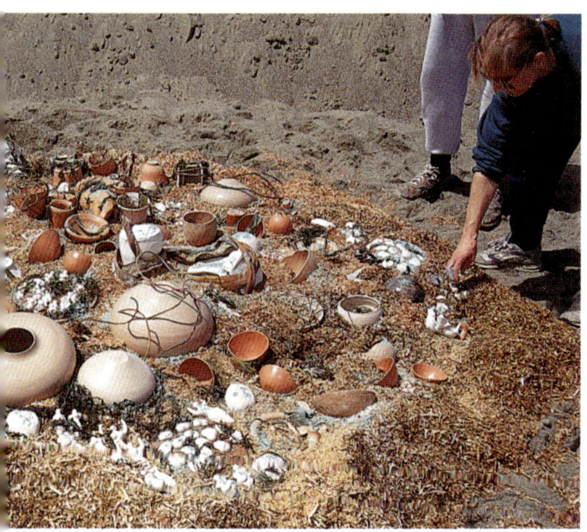

Carol Molly Prier loading a pit with wood shavings, pots, seaweed and copper carbonate.

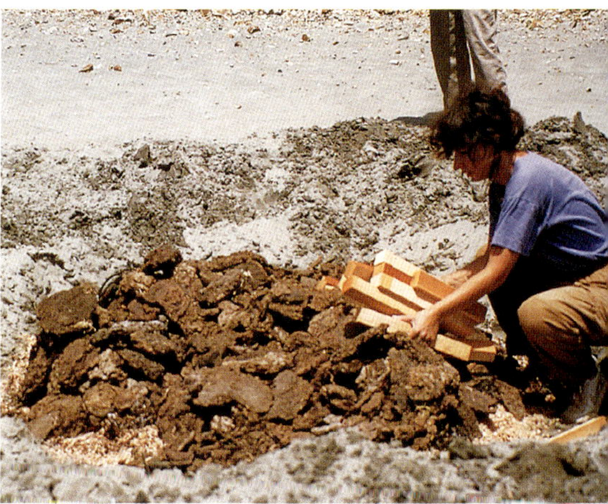

The pit is filled with generous quantities of dried manure and scrap wood, then lit. After burning the dung leaves a thick blanket of ash, keeping the pots from cooling too quickly.

and the slight slope encourages a draught through the pit.

Jane Burton fires in a variety of locations, digging her firing pits in round, square or rectangular configurations, depending on the number of pieces to be fired. However, she finds that the rectangular shape is the most fuel-efficient. The depth varies from 1–2 m (3–7 ft), and if digging in sand she may line the sides of the pit with sheet metal to keep it from collapsing. She starts with a layer of sawdust 15–30 cm (6–12 in.) deep, placing bisque-fired pots in a single layer with space between them for the salt and copper oxides required for colour. She uses dung when available, in the form of dried cowpats, and wood, including scrap timber and split oak firewood. The dung helps to insulate the pots from the wood as the fire heats up, and from the cool air as the fire dies down. She may cover the pit loosely with sheets of corrugated metal after 20 minutes or so to achieve a hotter fire with less wood, though an uncovered pit will burn faster.

Carol Molly Prier began pit-firing in 1975, after studying the methods of Native American potters who fire in the open, directly on the ground rather than in a pit. She started out firing that way, but began digging a shallow pit to protect the pots from wind. Firing on a sandy Californian beach, she digs a round pit about 2.4 m (8 ft) in diameter and between 46 cm (18 in.) and 60 cm (2 ft) deep. She puts about 10 cm (4 in.) of sawdust in the bottom, leaving room to walk around the pots inside the pit.

Since Prier doesn't use terra sigillata, always burnishing her pots with a stone, the colour of the clay body becomes the underlying colour of the finished pot. Sometimes she uses a smooth white earthenware, and for richer colours she uses a finely grogged buff stoneware clay. She has pit-fired her own work raw, without a preliminary bisque-firing, but firing raw pots so rapidly risks breakage.

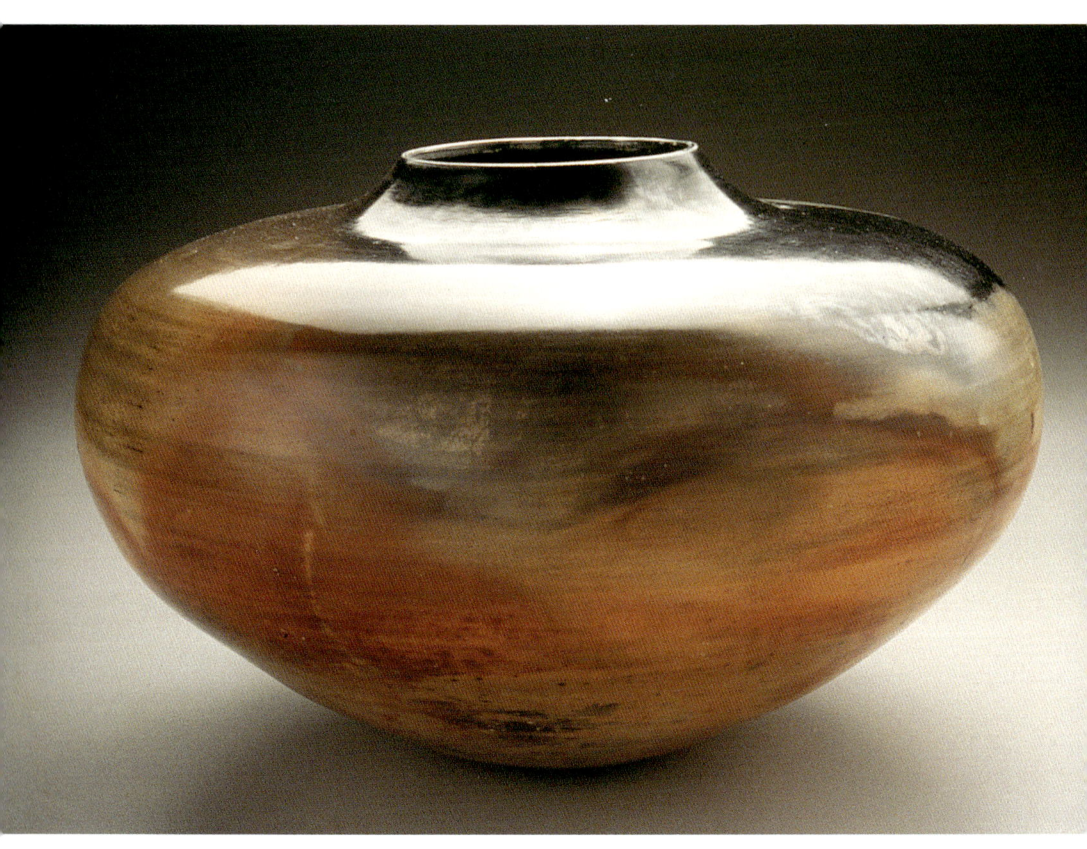

Vessel by Carol Molly Prier, 2007. Ht: 23 cm (9 in.), w: 33 cm (13 in.). Burnished and pit-fired. *Photo by Charlie Frizzell.*

Therefore when she conducts a workshop she always bisques to cone 010 before pit-firing. The pieces are placed in a single layer, not piled on top of each other, then seaweed and copper carbonate are sprinkled around them. Dried dung and wood are piled on top of the pots, filling the pit completely. For a 2.4 m (8 ft) diameter pit she uses about four dustbin-size bags of dung. She uses a mixture of wood, avoiding plywood, particle board, painted wood or wood with any nails. A pit this size holds enough cow dung and wood to generate the heat required for good colour. The art of firing this way is making sure not to get everything too hot, which can 'burn off' the burnish. If the wind causes the heat to rise too quickly it can cause pots to crack, or cause cracks in the burnish, so on windy days she sets up sheets of corrugated plastic to block the wind.

If you're beginning to think this prescription is out of your reach – sandy beach, wind, a pit 1 m (3 ft) deep and 2.4 m (8 ft) in diameter, lorry loads of manure and firewood – don't despair. It is possible to get good results with a much smaller effort. One solution is to add an air supply as a substitute for the wind and natural draught of a long pit. Firing in Colorado, at high altitude where there is less oxygen, in a small pit only 1 x 2 m (3 x 6 ft), I found it difficult to get a fire hot enough to coax colour onto the pots. Pondering about the two ingredients you need to keep a fire going – fuel and air – I decided that since I had enough fuel, I needed to get more air into my pit. Therefore I began to use a blower to force air into the pit after it was covered. This did the trick, getting the pit to temperatures around 760°C (1400°F) (measured by putting a pyrometer into the pit).

My pit-firing takes place in a pit that is about 46 cm (18 in.) deep and 1m (3 ft) wide by 2 m (6 ft) long. Along the bottom of the pit are one or two pipes (galvanized steel ventilation pipes) which are perforated along both sides. Using a blower from a raku kiln, air can be pumped through the pipe or pipes and into the bottom of the pit. The pit is filled with 10–13 cm (4–5 in.) of wood shavings (pine shavings, sold as pet bedding, or shavings collected from a woodyard) almost covering the pipes, then with another inch or so of dried, used horse bedding – this is both a free source of wood shavings and a source of natural salts to help provide colour in the pit, but it isn't absolutely indispensable. It is possible to substitute shavings soaked in salt water and then dried for the horse bedding, if firing in an area where horse bedding is hard to come by – or simply to use clean shavings and lots of table salt.

Another free source of colour is dried used coffee grounds, which are high in potassium. Potassium and sodium play similar roles in glaze formulation, and they seem to also act similarly to form colour in pit-firing. Potassium in refined form, such as pearl ash, could be used in pit-firing, but it is far more

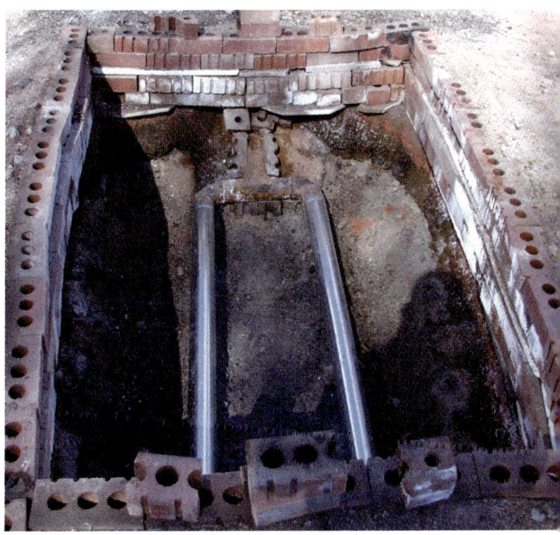

The pit, ready to load. The two pipes are pierced with holes along both sides to allow air to be blown into the pit during the firing.

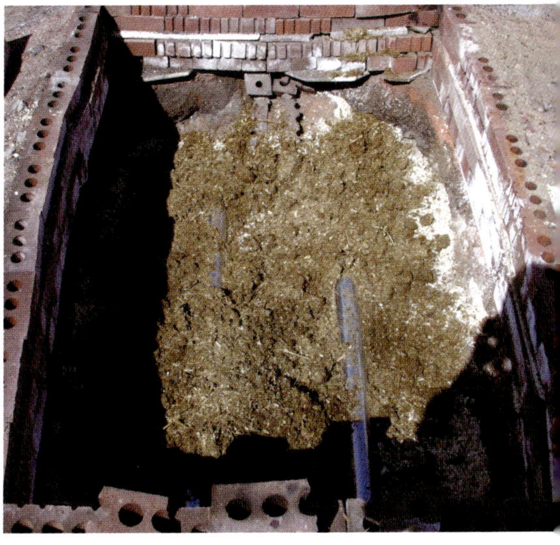

The bottom of the pit is covered with wood shavings almost to the top of the pipes. On top of the wood shavings several boxes of used shavings from a horse stall help to contribute minerals. A couple of clay boxes full of coffee grounds will be poured into the pit next, then 225–450 g (8–16 oz) or so of salt will be sprinkled on and lightly stirred in.

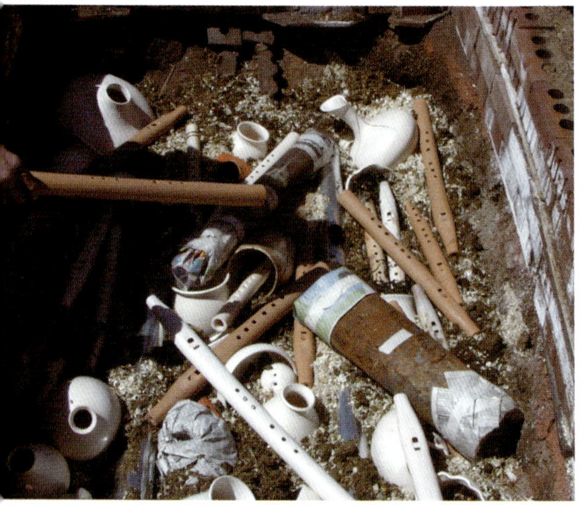

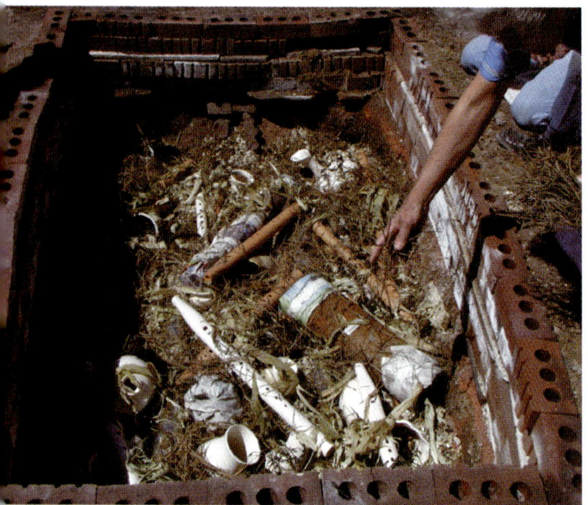

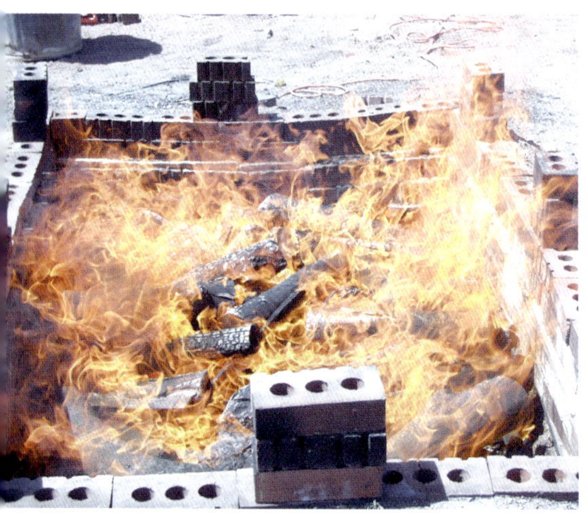

expensive than table salt – however, any organic material containing potassium is worth trying in a pit. In an urban area used coffee grounds in mass quantities are much easier to acquire than used horse bedding: simply ask your local café or coffee shop to save them for a day or two.

Onto this fragrant bed is sprinkled up to about 450 g (1 lb) of table salt which is lightly stirred into the top layer of the bed. Salt is your work-horse colour source in a pit. If you couldn't find horse bedding or coffee grounds, the salt alone would ensure colour on your pots. How much salt you use depends on how big your pit is and how many other sources of colour you have, and it takes trial and error to get right. You don't really want too much salt, as un-vaporised salt can melt onto your pot and form crusty spots – and you should do your utmost not to breathe any fumes, either. Another relatively inexpensive colourant to try is ferric sulphate, which is sold as fertilizer. Other gardening supplies to try include copper sulphate and Miracle-Gro.

Once the shavings are in place, the pots are nestled into the prepared bed.

LEFT, FROM TOP Pots, including flutes and a loaded saggar (the piece of pipe in the lower right) are loaded between the air pipes, mostly in a single layer. Some pots are wrapped in salt-soaked materials, with newspaper taped around them to hold everything in place.

Salt-soaked pine needles are distributed around and on top of the pots. Salt-soaked corn husks are also added, as well as seaweed if available. Hardwood scraps, aspen logs and newspaper are then piled on top of the pots and lit.

The wood has begun burning evenly across the pit.

They can touch, and I may even pile them on top of each other. Where the pots are partially buried in the shavings, they will turn black. Now more sources of colour can be added. Copper or steel wire wool or mesh (such as scrubbing pads for washing dishes) will leave distinct black marks, sometimes with rust-coloured or red clouds nearby. Banana peel, another source of potassium, may leave pinkish marks.

Seaweed is a popular addition to a pit; it is another source of salt which can be gathered in large quantities for free if you happen to live near the sea or a salt marsh. In land-locked Colorado, seaweed is hard to come by, so as a substitute I soak various vegetable materials in salt-water. String, craft supplies such as excelsior (shredded wood) and sphagnum moss, pine needles, corn husks, iris leaves, rags – anything organic which can be soaked in salt water and dried out is worth trying. These materials are tucked between pots and sprinkled on top of them. It is also possible to position some of these materials around pots before

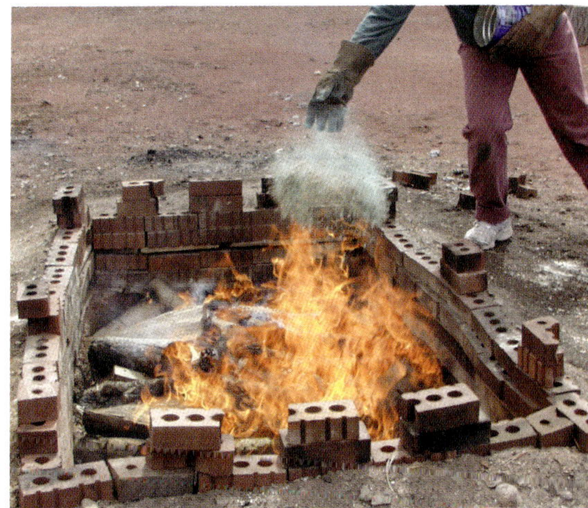

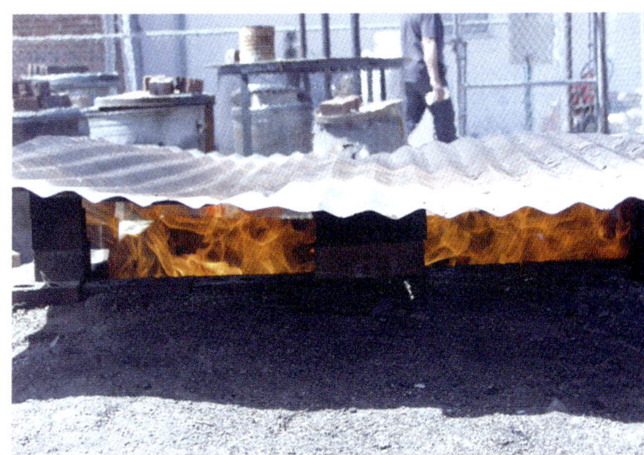

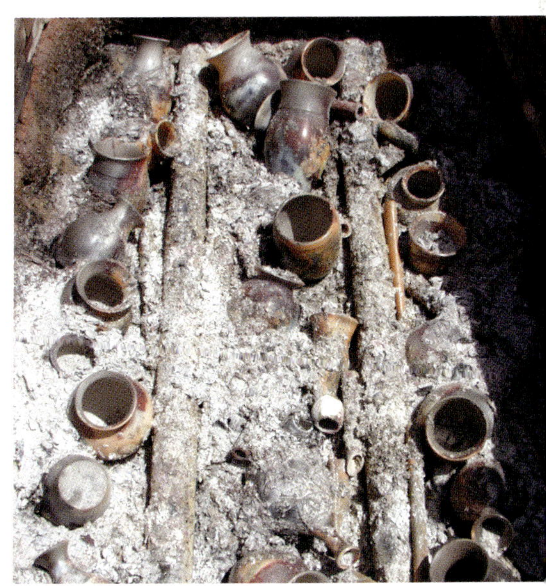

RIGHT, FROM TOP Once the wood has begun burning across the pit, handfuls of wood shavings mixed with copper carbonate are tossed on to the flames. After the copper carb has been added the pit is covered with sheets of galvanized metal to keep the heat in.

The covered pit. Notice that the cover rests on stacks of bricks to allow some air passage.

The next day the cooled pit is uncovered to reveal colourful pots.

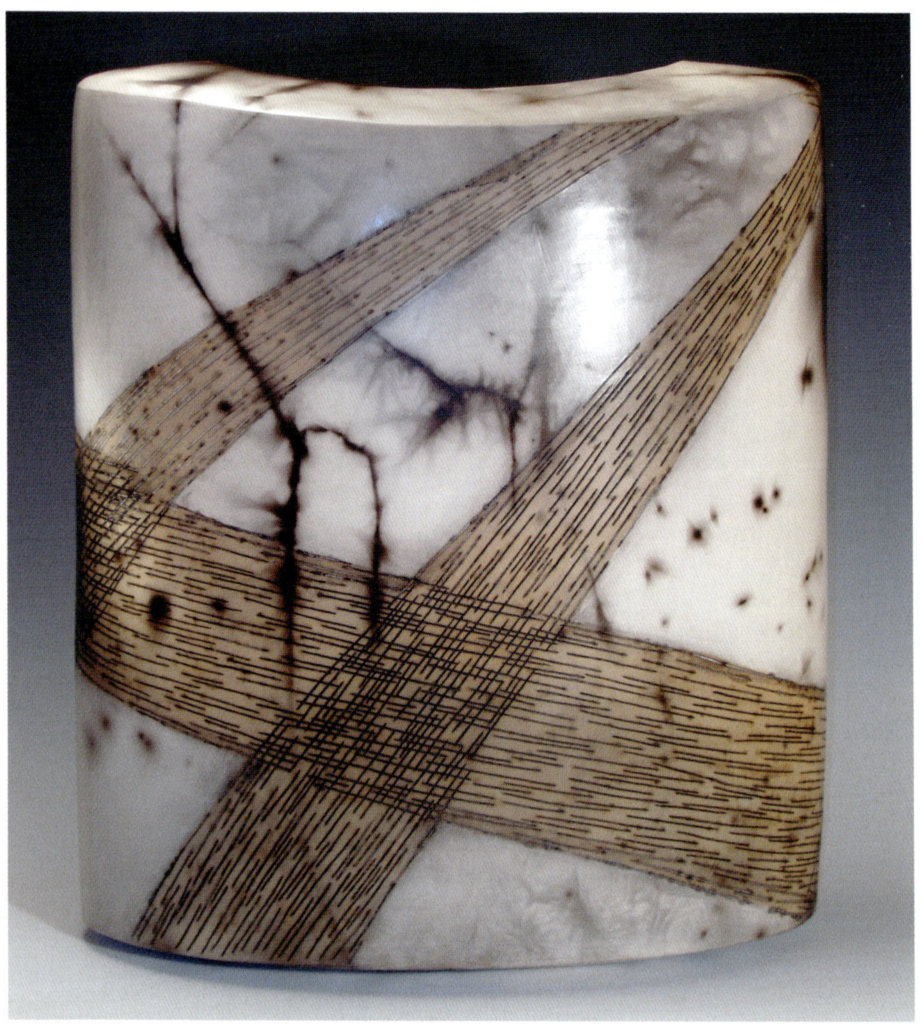

Eliptical Pot by Antonia Salmon, Ht: 60 cm (23½ in.). Hand-built from white stoneware clay, burnished using teaspoon and plastic rods at least four times. Bisque-fired to 1000°C (1832°F) and then smoke-fired several times. Polished with wax after firing.

placing them in the pit, wrapping them in newspaper to hold everything in place. Small pots may be placed inside larger pots, with a generous handful of shavings and some of the salt-soaked materials. Now you can cover the pots with whatever fuel you can find in sufficient quantity. Cabinet shops are a good source of kiln-dried hardwood or pine scraps, and often woodyards will sell or give away timber off-cuts. In Colorado I use aspen wood (dried for at least a year) in addition to hardwood scraps: because aspen trees don't have a long lifespan, they are small and easy to harvest and cut to size, and the wood burns quite hot and fast.

Once the fire is lit and burning evenly,

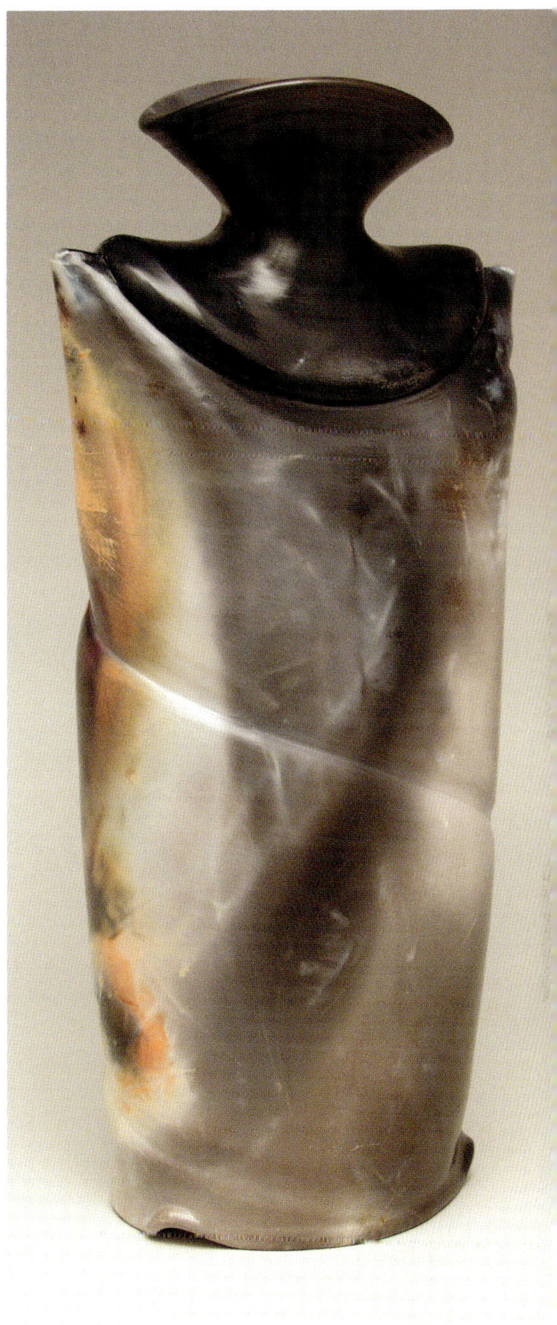

RIGHT *Für Elise* by Sumi von Dassow, 2004. 35 x 15 x 15 cm (14 x 6 x 6 in.). Wheel-thrown and altered from B-Mix clay (white stoneware); terra sigillata sprayed on and burnished with a chamois-leather; bisque-fired to cone 010; pit-fired with salt, coffee grounds, copper carbonate and salt-soaked pine needles. The pine needles made the whitish marks on the front of the piece – instead of colouring it, they resisted the smoke because this part of the pot was in a cooler area of the pit. *Photo by Sumi von Dassow.*

I add copper in the form of copper carbonate mixed with wood shavings (1/4 cup or so in half a coffee tin of shavings), which is sprinkled directly onto the flames. A related technique, from potter Dean McRaine, is to blow copper carbonate into the fire with compressed air. The challenge is to make sure the fine powder is dispersed evenly and doesn't land in large quantity on individual pots, because it will make black splotches instead of red fuming where it comes into direct contact with hot pots.

After adding the copper carbonate, I cover the pit loosely with sheets of galvanized metal and turn on the blower to add air from below. I don't ever restoke the fire, but let it burn for 4 or 5 hours with the blower on and the cover in place. At this point the wood has burned down to coals and ashes, and the colours on the pots can be seen if the cover is lifted. I turn the blower off, cover the pit more tightly, and let it smoulder and cool overnight before unloading.

Chapter 6

Saggar firing

Saggar firing has evolved as a way to achieve a result similar to pit-firing, but one pot or one kiln-load at a time. A saggar is a lidded container which is used to contain and isolate a pot during its firing. Though saggars were originally invented to protect pots from harmful fumes inside the kiln, such as sulphur fumes in a coal-fired kiln, today their use is exactly the reverse: saggars are used to contain fumes around a pot to allow the pot to pick up colour. The requirements for saggar firing are minimal compared to pit-firing – a kiln, preferably gas rather than electric, is all you need. A raku kiln is fine for one or two pots.

The same ingredients that go into a pit-firing also go into a saggar: sawdust or wood shavings, salt, copper carbonate, and perhaps copper wire or steel wool. A little bit of each material is placed inside the saggar along with a pot, then the pot is fired to somewhere between 760 and 987°C (1400 and 1800°F). That's about all there is to it.

First, of course, you need a saggar. Make it from a coarse sculpture clay that will resist repeated heating and cooling – many suppliers formulate a clay specifically for saggars. Charles and Linda Riggs, who have conducted numerous saggar-firing workshops, suggest a clay such as Laguna's 548 or Soldate 60. The shape and size of the saggar depends on the pot being fired, but most potters will throw a pair of deep bowls, a larger one to put the pot inside, and a small shallow one for a lid. The lid doesn't need to fit tightly, unless you want black pots. The Riggs suggest making these bowls 1–1.5 cm (3/8 – 1/2 in.) thick and 10 cm (4 in.) higher and wider than the pot which will go inside it. Saggars should be bisque-fired to cone 06. It is convenient to make several saggars of different sizes at one time, so that when you are ready to fire you can select a suitable pair for the size of pot to be fired. Saggars can be reused, though they will break after a few firings so you'll have to keep making new ones.

The process of preparing a saggar for firing is quick and simple. A burnished pot, bisque-fired to cone 010 to cone 08, is placed inside the saggar on top of a support such a a chunk of firebrick or broken kiln shelf, to isolate it from direct contact with chemicals. A handful or two of wood shavings is placed around the pot in the bottom of the saggar, then 1–2 tablespoons of salt and/or copper carbonate are sprinkled around the bottom edges of the saggar, away from contact with the pot. Steel wool or copper wire can be draped over the pot before the lid is placed on the saggar.

Linda Dadisman uses this method to fire her graceful long-necked bottles. She uses a small wad of ceramic fibre or steel wool between the rim of the

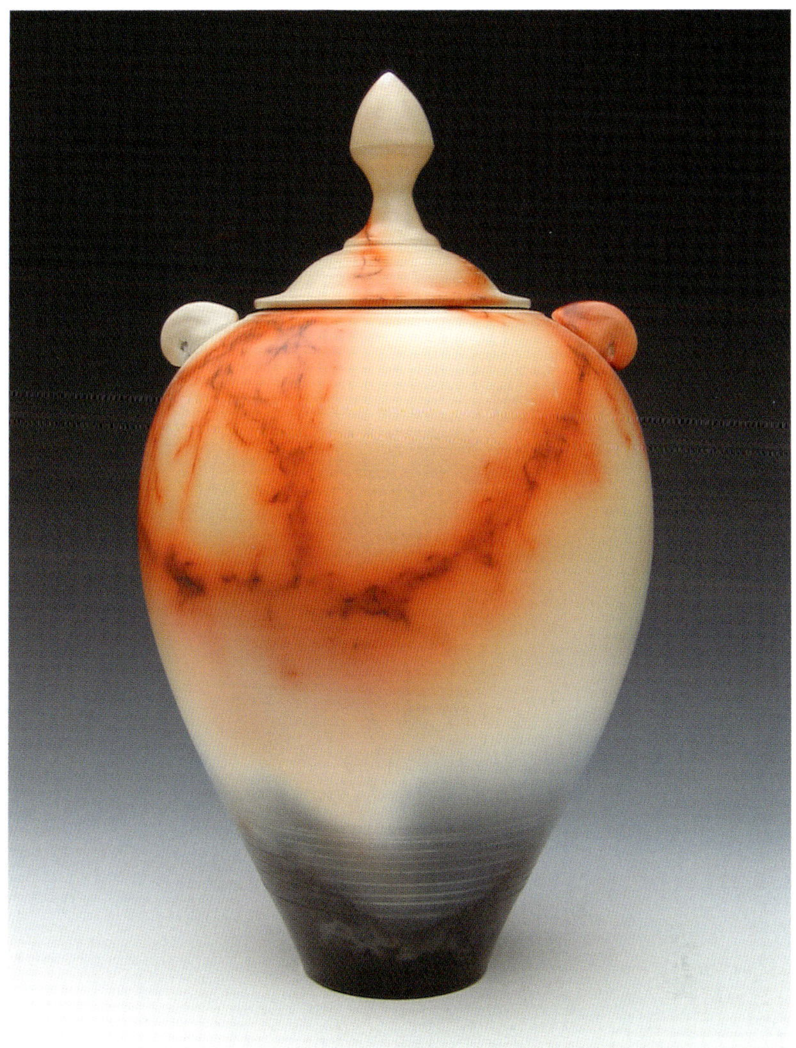

Lidded vessel by Charles and Linda Riggs, 2003. Ht: 30 cm (12 in.), w: 20 cm (8 in.).White stoneware sprayed with white terra sigillata and polished with a soft cloth; bisque-fired to 010; saggar-fired in a raku kiln with wood shavings, steel wool, copper and salt.

saggar and the saggar lid, to allow enough airflow during the firing to keep the pots from turning completely black. She sometimes uses Miracle-Gro pellets inside the saggar as another source of colour, or sea salt instead of table salt for a slightly different colour. She also suggests using wood shavings that have been soaked in ferric chloride and then dried, to give a rust colour. A favourite technique of hers is to pull a section of copper dish scrubber over the pot like a sleeve, where it will leave a distinct pattern. The copper wire is light-weight enough to be easily stretched and torn for a more random pattern.

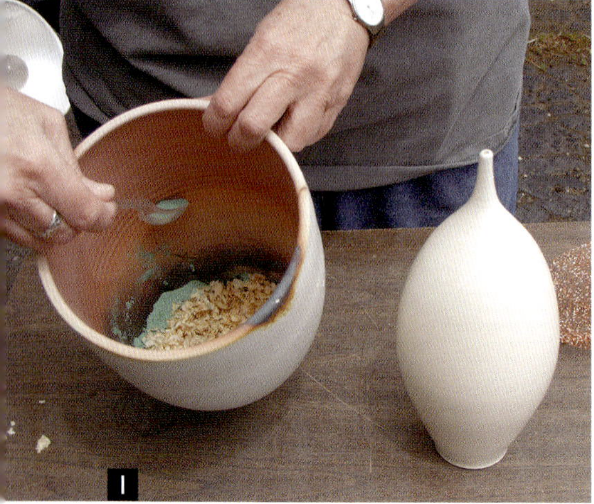

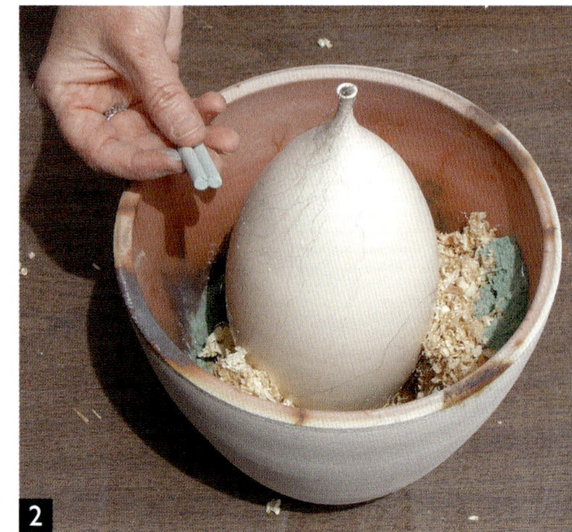

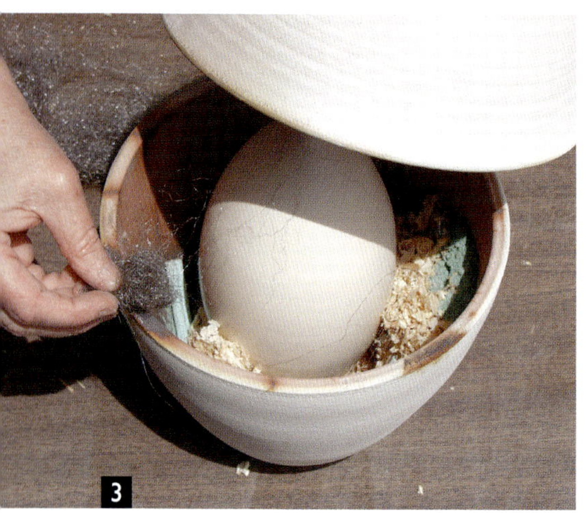

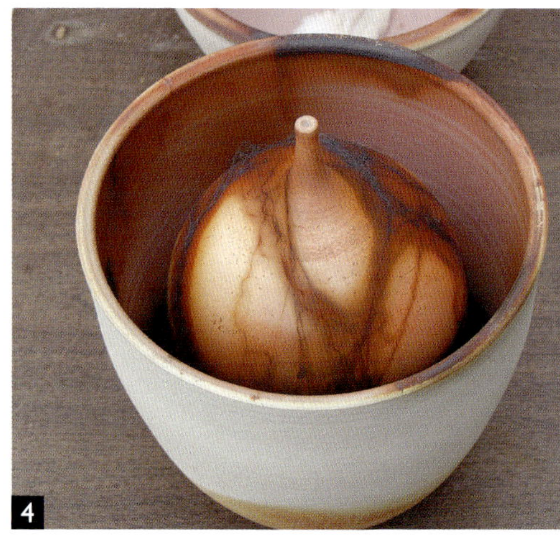

1 Linda Dadisman begins loading a saggar with a handful of wood shavings and a spoonful of copper carbonate – notice the copper carbonate is carefully placed where it won't directly touch the pot.

2 A few strands of copper wire, a torn piece of copper scrubbing pad or, in this case, a few strands of steel wool can be wrapped around or draped over the pot. The pot is placed in the saggar and Linda places a stick of Miracle-Gro fertilizer at the edge of the wood shavings, where it won't directly touch the pot.

3 Linda has added another handful of wood shavings and another spoonful of copper carbonate, and now she is placing the lid on the saggar. She uses a piece of steel wool to prop up the lid so air can get in during the firing. Alternatively, she may just use a bit of Kaowool (ceramic fibre).

4 The saggar after firing. A very few strands of steel wool have made a great deal of colour on the pot.

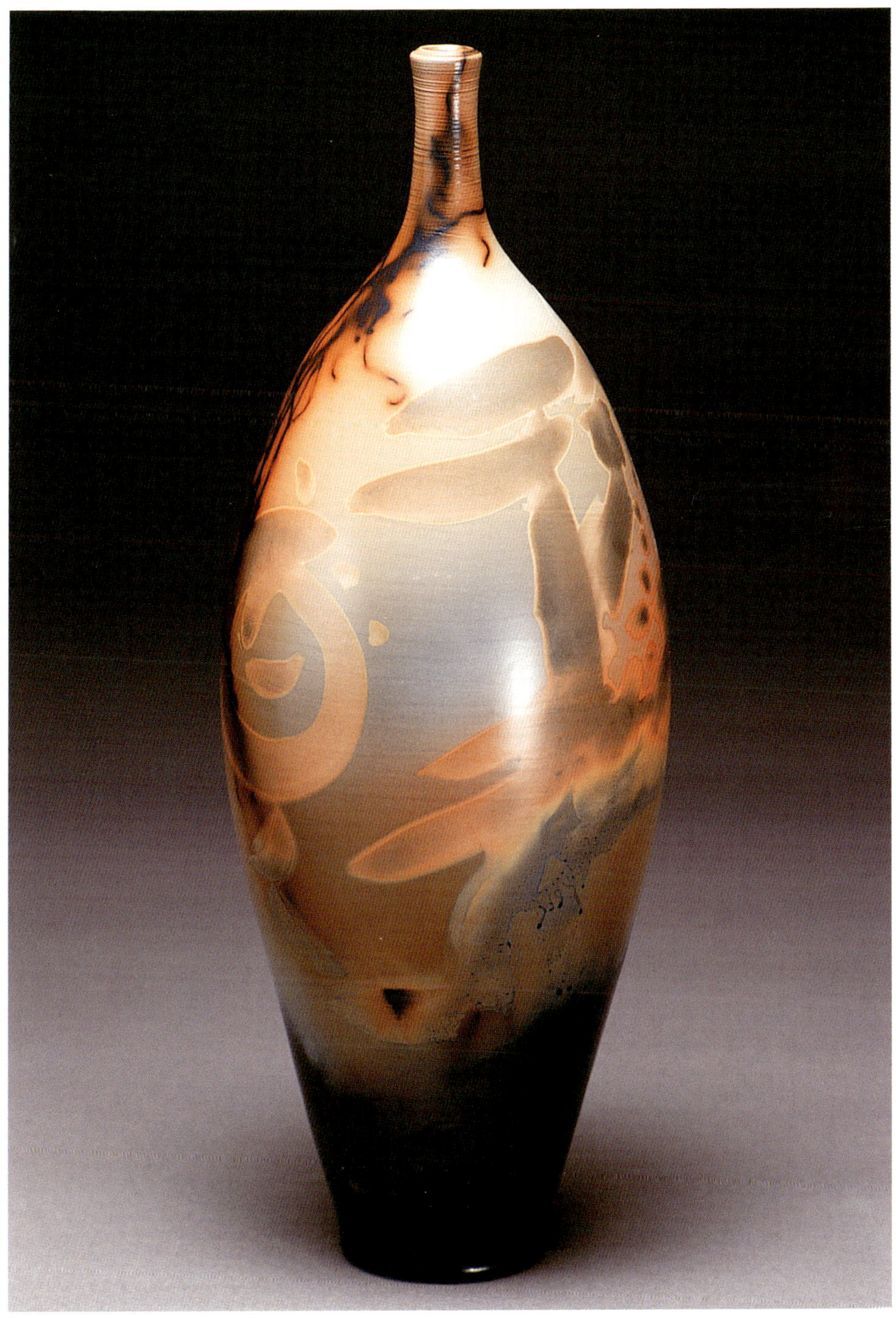

Bottle by Linda Dadisman, 2007. Ht: 30 cm (12 in.), w: 13 cm (5 in.). Saggar-fired with salt, copper carb, steel wool, pet bedding and ferric chloride.

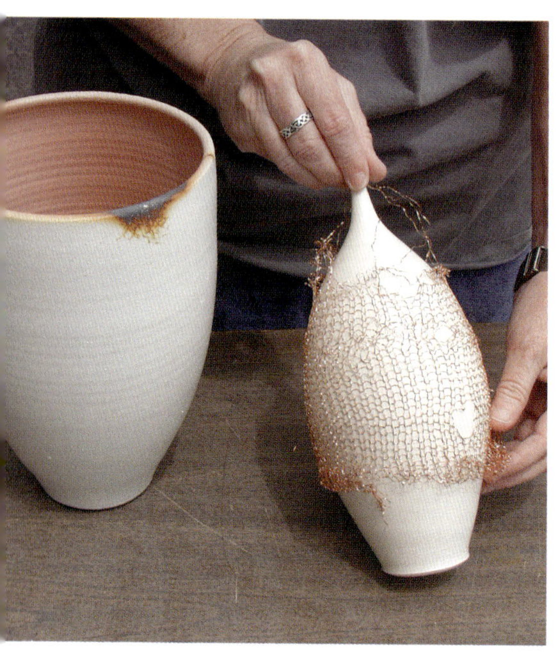

Instead of copper wire, a section of copper scrub pad can be placed around a pot like a sleeve. This material is easy to pull apart and the mesh can be left relatively intact, as in this photo, or artfully torn and frayed.

Her firing cycle is to bring the kiln up to about 880°C (1620°F), which in a raku kiln takes 20–30 minutes, and hold it at that temperature for 20 minutes or so. It is unloaded after it has cooled completely, to avoid damage to the pots from rapid forced cooling.

Joan Carcia uses salt, green hay, salt marsh hay, seaweed, straw and sawdust inside her saggars, mixing iron oxide, copper carbonate and cobalt carbonate with some of the salt marsh hay. She

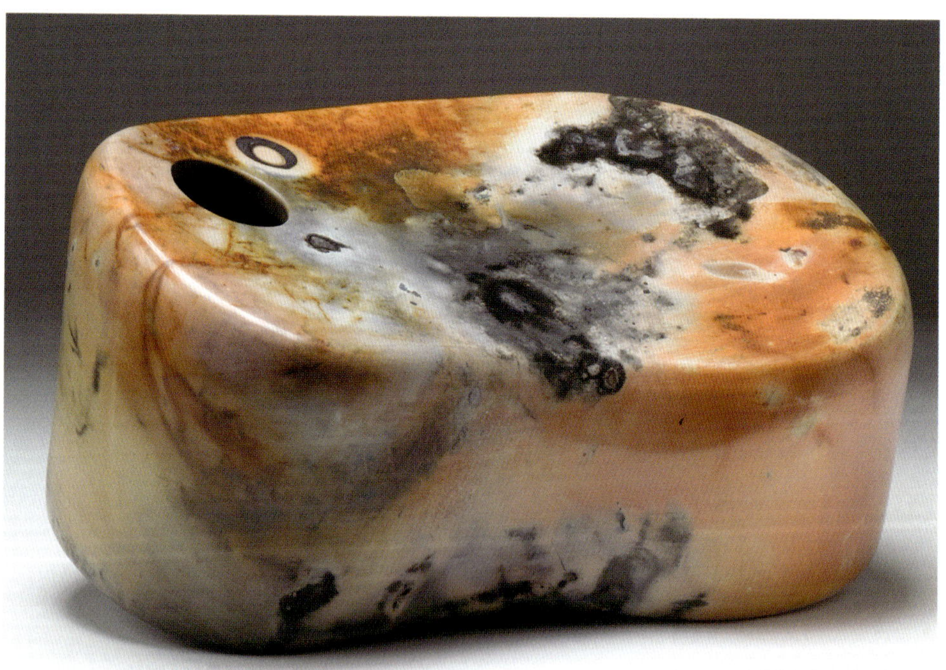

Sculpture by Joan Carcia, 2006. 12 x 24 x 14 cm (4.5 x 9.5 x 5.5 in.). Hand-built from slabs using low-fire white clay, burnished with a stone; saggar-fired with green hay, salt, seaweed, copper, salt marsh hay with iron oxide and cobalt carbonate. *Photo by Steven Gyurina.*

covers the saggar with a tile or piece of kiln shelf and fires in a gas kiln to cone 012. She fires very slowly, especially during the first few hours, allowing the kiln to take six or seven hours to reach temperature. This gives the pots plenty of time to soak in chemical fumes, ensuring a good absorption of colour into the clay body.

Edge Barnes uses one part each of copper sulfate, fine sea salt, cottonseed meal, baking soda, and ½ part each copper carbonate and titanium dioxide in his saggars. For larger pots he mixes these materials with water to create an evil-looking bubbling liquid he calls 'swamp juice' and brushes this juice directly onto the saggar to avoid having all the chemicals concentrated only near the bottom of the pot. He also uses coarse steel wool, copper wire and seaweed. For saggars for larger pots he suggests using two shallow bowls for the top and bottom of the saggar, and adjusting the height of the saggar by placing rings of thrown clay between the bowls. He punches holes in the rims of his saggars to allow airflow, and fires to 870°C (1600°F) in about an hour.

For a quick and easy variation on saggar firing, Barnes now prefers an aluminum foil 'saggar'. He paints each pot with ferric chloride (sold as etching solution for printed circuit boards) using a cheap foam brush, rotating it on an inexpensive banding wheel with a plastic top as he brushes it on. Other potters spray this material using an inexpensive spray gun. 'Inexpensive' is emphasized, as ferric chloride is caustic and toxic. It will ruin good brushes, eat away at metal parts on a spray gun, and corrode your metal banding wheel if it comes into contact with it. If you choose to spray ferric chloride, you must wear gloves, goggles and a face mask and spray in a well-ventilated area. Despite all these serious disclaimers, ferric chloride is fairly commonly used because it reliably yields spectacular pink to orange colours.

After all the pots are coated with the ferric chloride, Barnes mixes up the same swamp juice in a shallow bowl with just enough water to make the mixture froth. Once this has bubbled up and increased in volume he touches the pot to the bubbling mass. This leaves a lacy deposit on the surface where it contacts the pot. The swamp juice can also be brushed or splashed onto the pot. The saggar is made with foil that has been crinkled up and then spread back out. He scatters a little coarse steel wool, raw cotton and wood

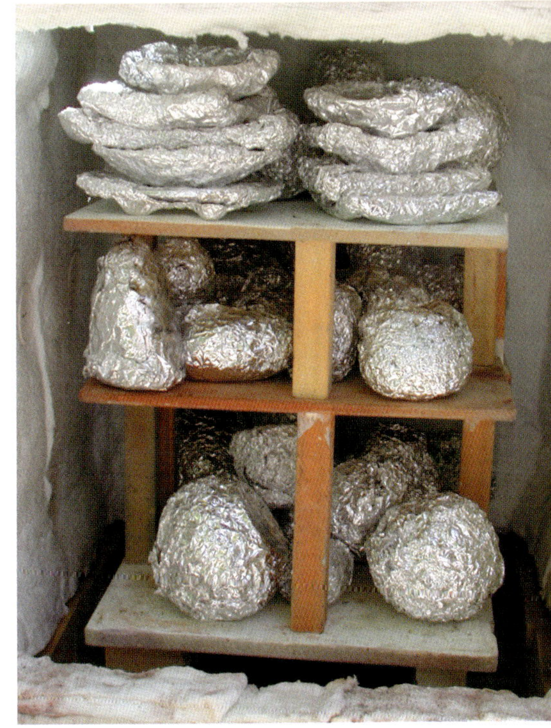

Edge Barnes's kiln loaded with tin-foil saggars, ready to fire. *Photo by Edgeworth Barnes.*

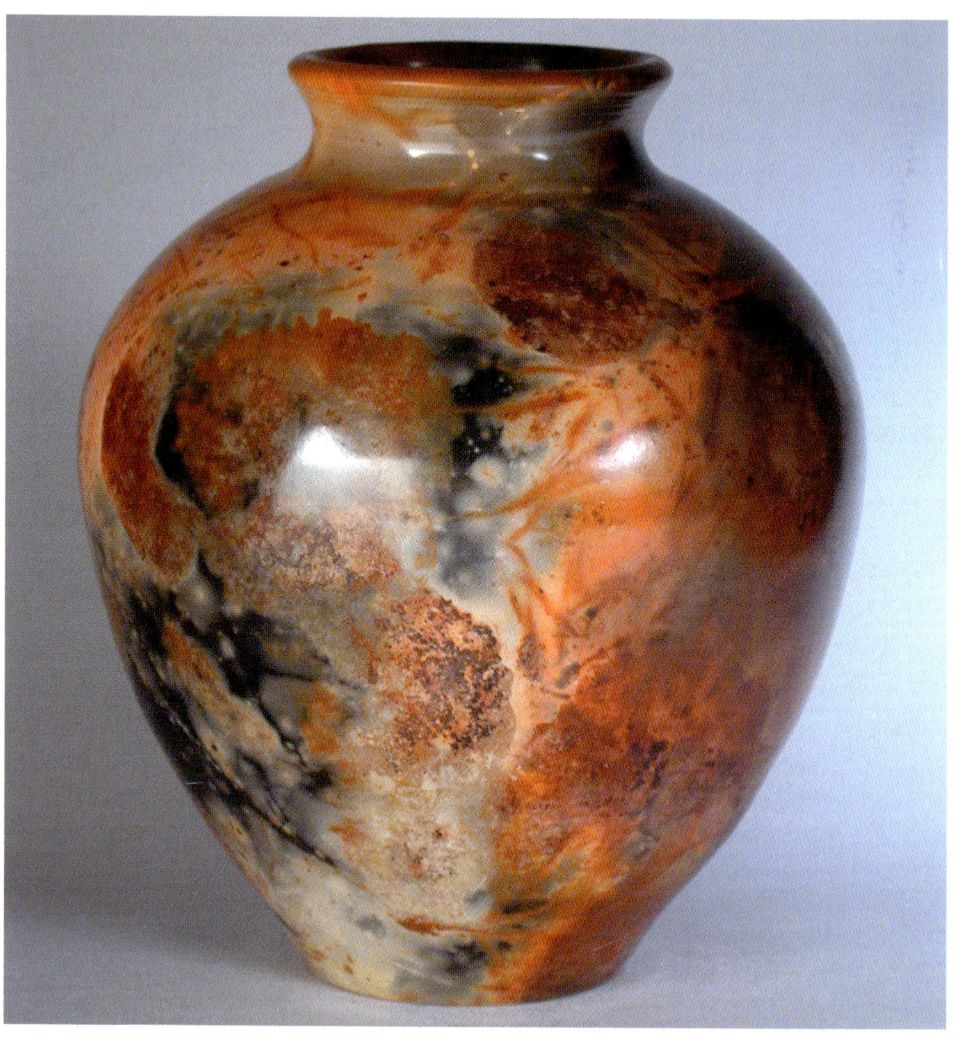

Seaweed Vase by Edgeworth Barnes. W: 17 cm (6½ in.), Ht: 20 cm (8 in.). Wheel-thrown, burnished with a stone, and fired in aluminum foil saggar.

chips on the foil. Next he places moistened seaweed over these materials. Copper wire or pieces of copper dish scrubber can also be added to the mix. Next the pot is placed, usually top down, onto all these items. More seaweed, cotton, wood chips and steel wool are then placed over the pot. Finally, the foil is wrapped around to cover the pot and pressed into close contact. The operative words here are 'a little' of each of these materials – too much combustible material can result in solid black pots if the foil doesn't burn away.

The pots are tumble-stacked in a kiln and fired to 680°C (1260°F) (about cone 017), at which point much of the foil will have vaporised. It is important to do this outside away from people and homes! Ferric chloride and the other materials will create very toxic smoke as they burn.

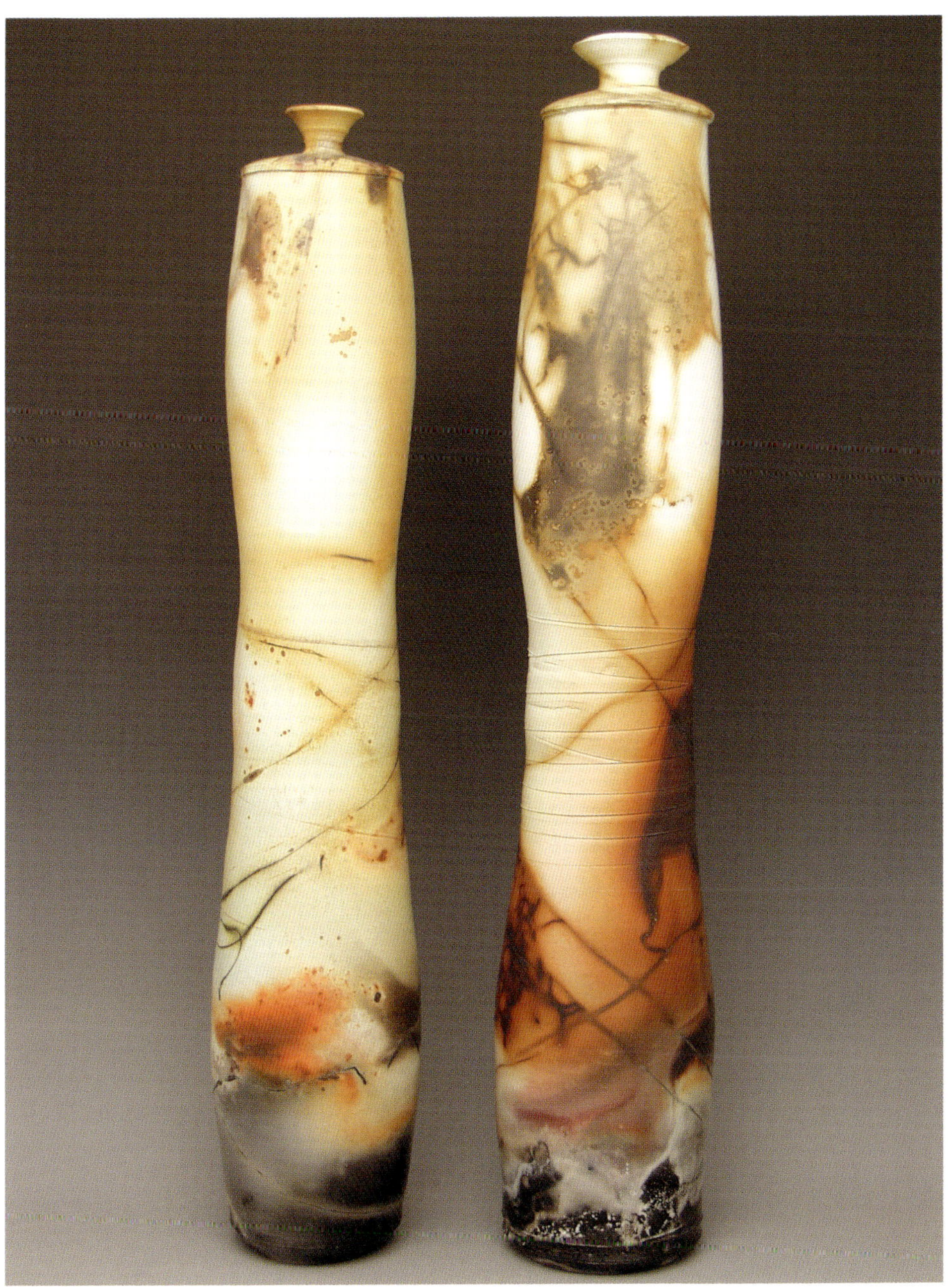

Undulates by Bob Smith. Ht: 70 cm and 74 cm (27.5 in. and 29 in.). 2007/2008. Made from Ash clay (Mile-Hi Ceramics), with OM-4 terra sigillata. Bisque-fired to cone 08 and individually saggar-fired in a three-storey saggar consisting of a ring between two tall bowls. Colour from salt, copper carbonate, fine wire and steel wool. Taller vessel was partially wrapped in dampened sushi nori (seaweed) towards the top.

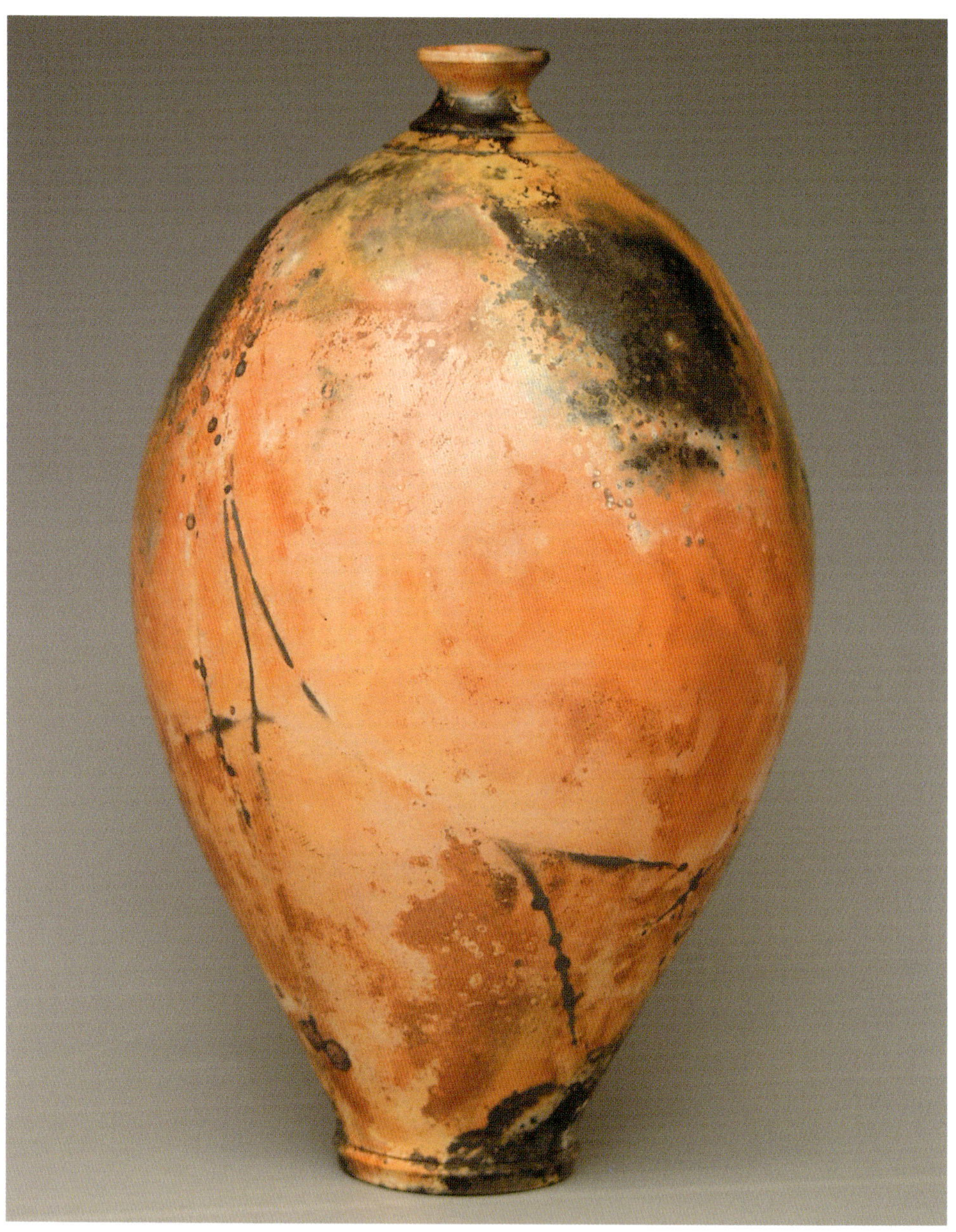

Vessel, by Bob Smith. Ht: 28 cm (11 in.), dia: 15 cm (6 in.), 2008. Fired in a tin foil saggar. 'I brush on varying amounts of ferric chloride (bought at Radio Shack* as Etchant Solution) onto the pot before wrapping the pot and other ingredients (gerbil bedding, salt, copper carbonate, and copper wire) in foil. I use two layers of heavy duty foil. I fire to around 600–640°C (1120–1180°F), and when the foil disintegrates, pull the piece, and try to spray it gently with water while I peel away the remaining foil.'

* *Radio Shack is an electronics store in the USA, other electronics stores may sell Etchant Solution.*

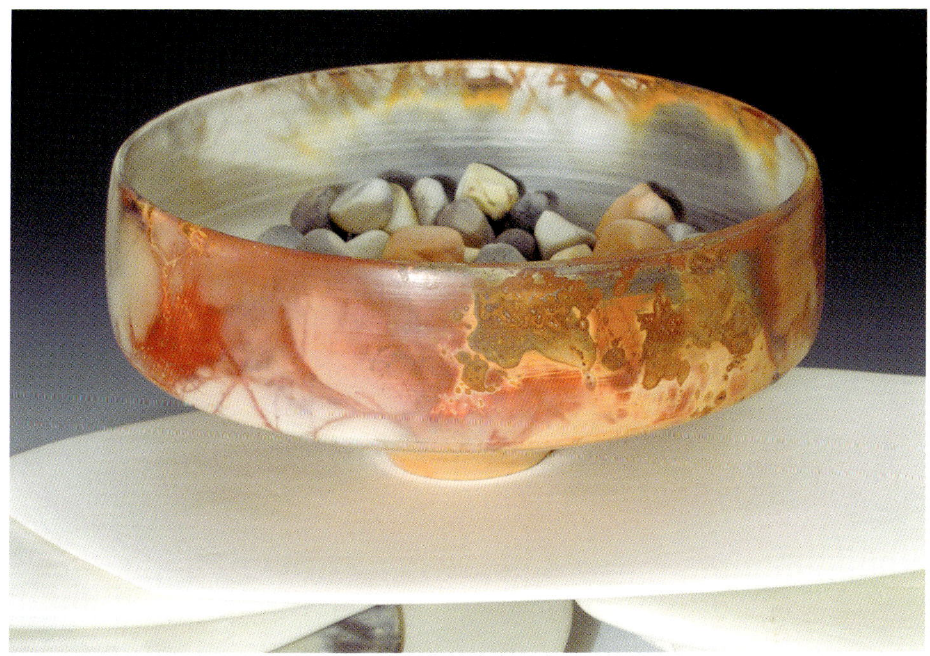

Collecting Bowl with Stones by Judith Motzkin, 2007. Ht: 38 cm (15 in.), w: 61 cm (24 in.), dia: 30 cm (12 in.). White earthenware, burnished with a rib; terra sigillata brushed on using a banding wheel; bisque-fired to cone 09, and saggar-fired in a gas kiln. *Photo by Judith Motzkin.*

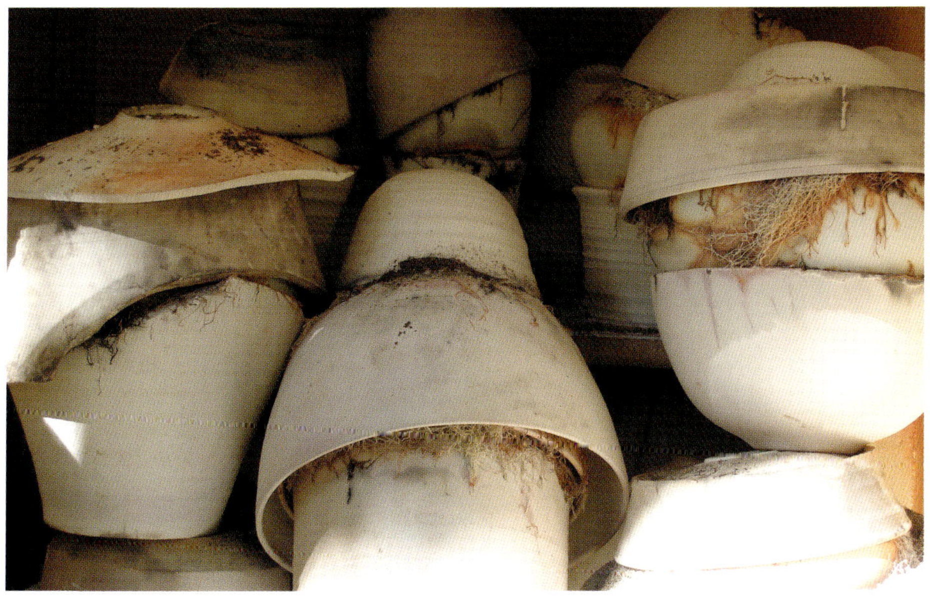

A tumble-stacked gas kiln with saggars before unloading. Notice the ashy strands of burnt straw poking out from the stack. *Photo by Judith Motzkin.*

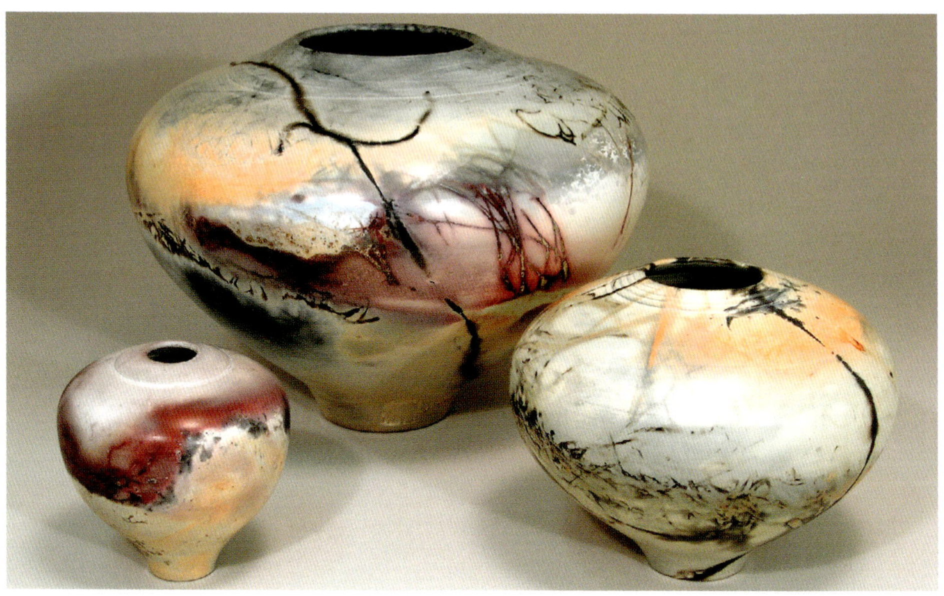

Three Vessels, by Judith Motzkin, 2007. Ht (tallest): 25.5 cm (10 in.). White earthenware, burnished with a rib, terra sigillata brushed on using a banding wheel; bisque-fired to cone 09 and saggar-fired in a gas kiln. *Photo by Judith Motzkin.*

I have used aluminum foil to saggar fire, with or without ferric chloride, using many of the same materials that may go into a pit. Often I will wrap a pot with toilet paper, then dab it all over with a sponge dipped in ferric chloride, salt-water, or pearl ash (potassium carbonate) dissolved in water. Any soluble form of sodium or potassium will give you a soft pinkish colour after firing. If you don't soak the pot too much, it can be fired immediately, without waiting for the solutions to dry. You can also get good results simply by wrapping the pot with seaweed or organic materials which have been soaked in salt – wood shavings, pine needles and corn husks are good options – then tightly wrapping the pot with foil. Copper or steel wire wrapped around the pot will leave a stark black line. The firing is to cone 017 in a well-ventilated kiln.

There are infinite variations on the theme of saggar firing. If you are black-firing, once you've found a method that makes your pots black there's no need to keep experimenting. With pit-firing, you are committing dozens of pots to each firing, so radical experimentation risks the loss of a great deal of work. In saggar firing, however, with a single pot in each saggar it is easy to try something different with every pot. If you are using a raku kiln you'll be firing one to three pots at a time, so you can see the results of each experiment before you try the next one. You can experiment with chemicals (copper carbonate or copper sulphate; salt, soda, and pearl ash; soluble or insoluble iron, manganese or cobalt sources) and organic materials such as seaweed, manure, straw or other combustibles soaked in salt water and dried; various types of metal wire,

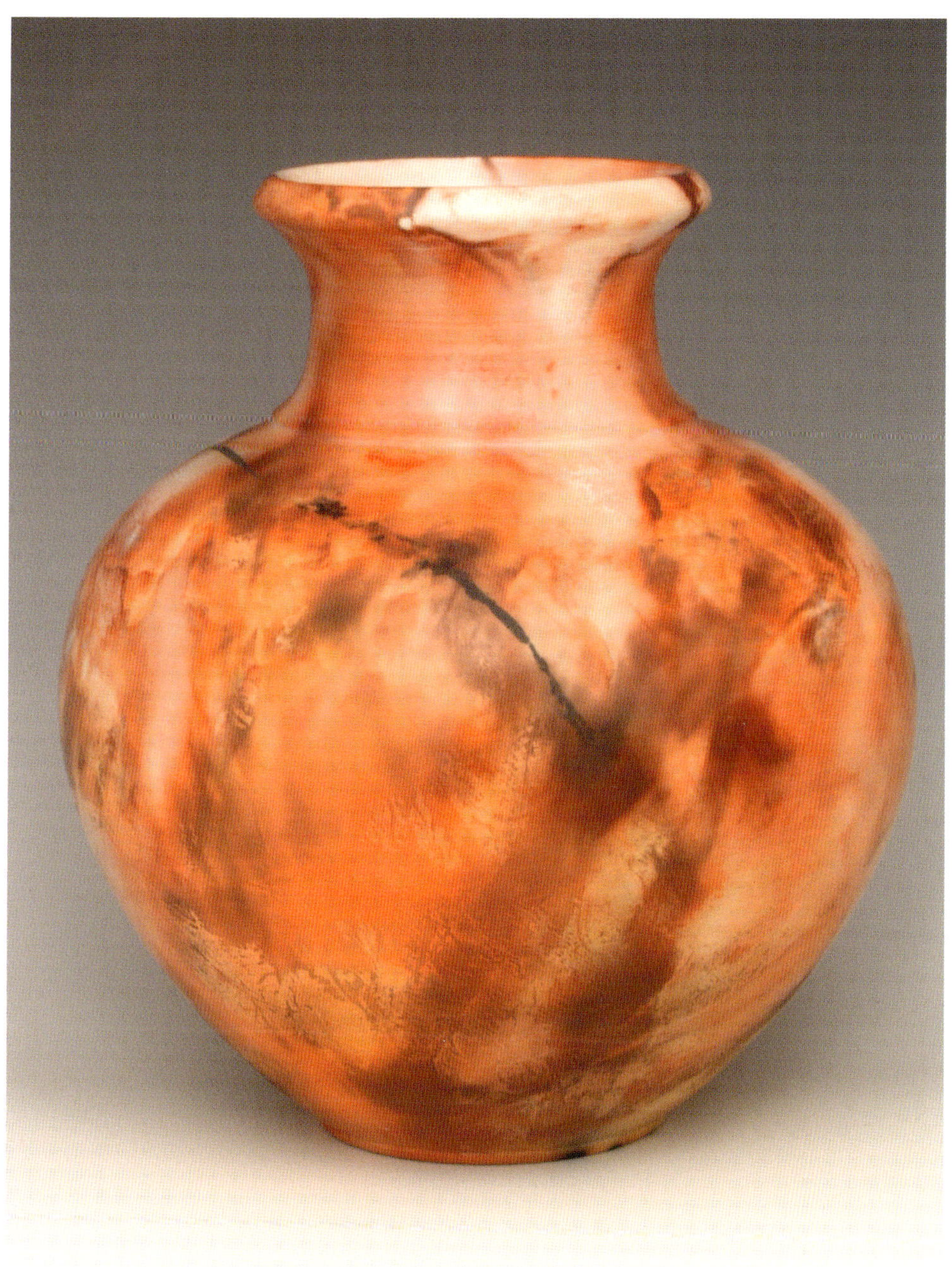

Vase by Sumi von Dassow. Ht: (25.5 cm (10 in.), dia: 23 cm (9 in.), 2005. Wheel-thrown from B-Mix clay (white stoneware), terra sigillata brushed on and burnished with a chamois, bisque-fired to cone 010. Wrapped in toilet paper, then dabbed all over with ferric chloride solution (wearing gloves!). A guitar string tied around the pot made the black mark on the shoulder. Saggar-fired wrapped in aluminum foil to cone 017. *Photo by Sumi von Dassow.*

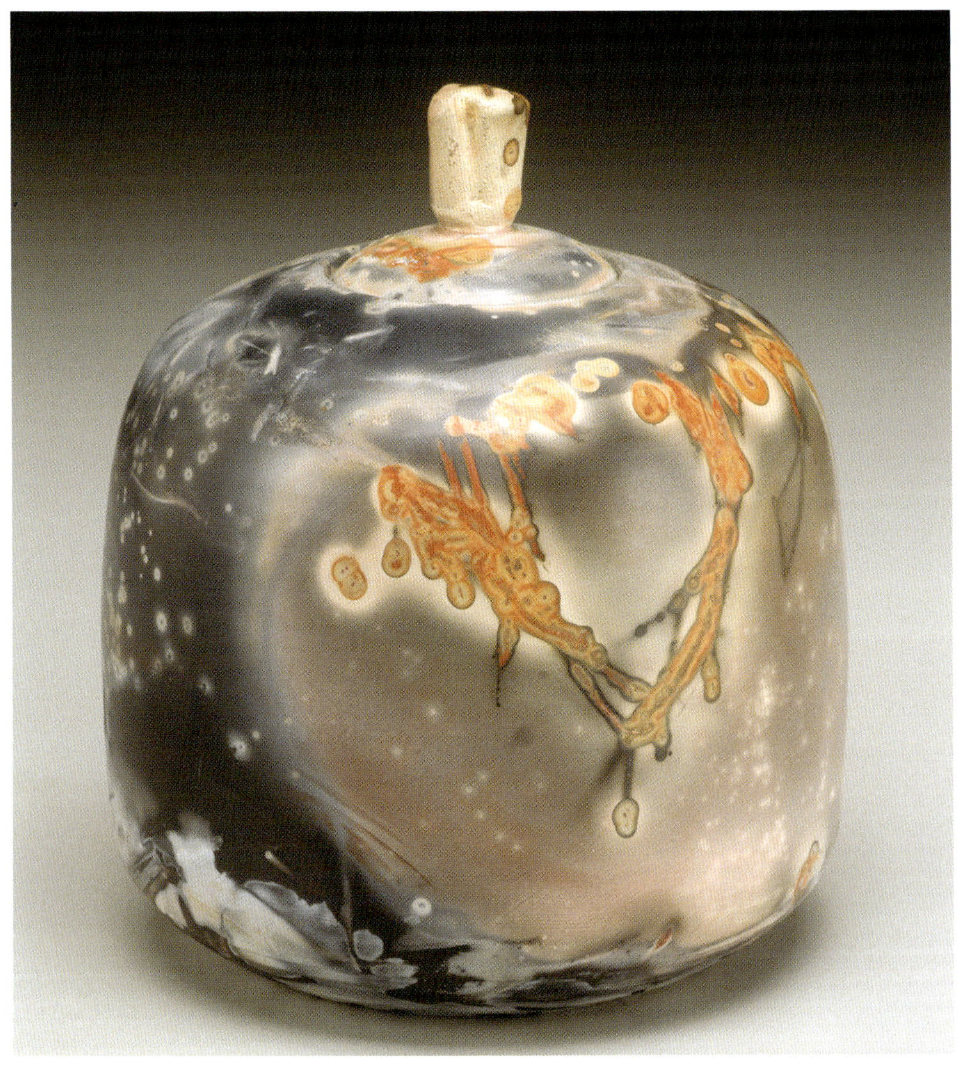

Vessel by Jan Lee, 2004. 23 cm (9 in.) x 20 cm (8 in.). Saggar-fired with sawdust, steel wool, salt, copper carb. and aluminum foil to cone 010 in clay saggar/fibre raku kiln.

shavings and wool; firing time and temperature; and different clay bodies and methods of burnishing. If you have a gas kiln, you can try loading several pots into a large saggar, or create a box of fire-bricks inside your kiln, to act as a large saggar for several pots, using kiln shelves for a lid. Or you can load many individual small saggars into the kiln at once, even stacking them for efficiency. It is the interplay of all these variables that makes each potter's saggar-fired work unique – and keeping careful notes as you experiment will allow you to reproduce favourable results – though Joan Carcia warns that, 'Just when you think, yes, I have it – this is great – when you try it again, it comes out totally different'.

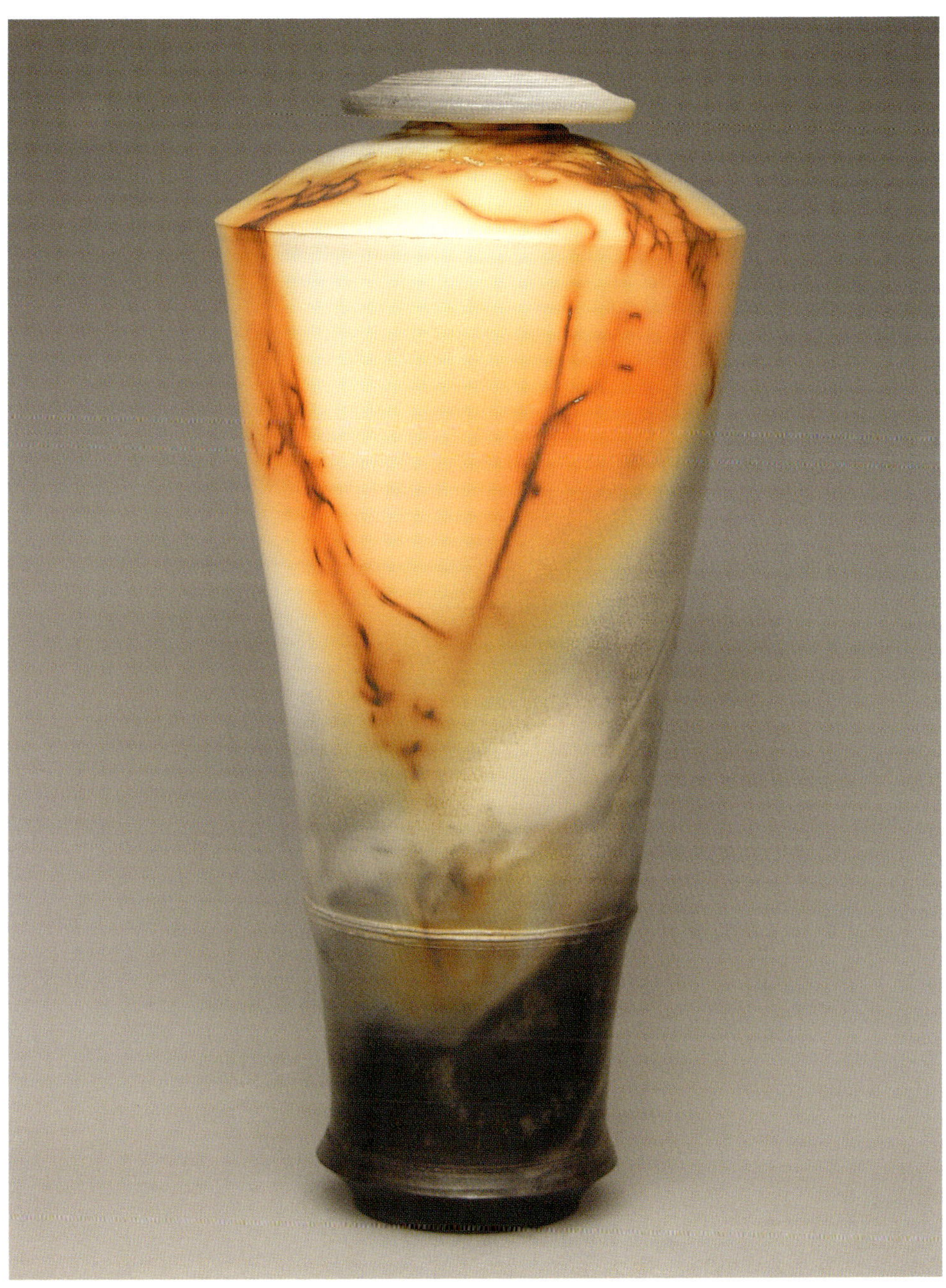

Saggar-fired vessel by Bob Smith. Ht: 33 cm (13 in.), dia:15 cm (6 in.), 2008. Made from Ash Clay (Mile-Hi Ceramics), with OM-4 terra sigillata. Bisque fired to cone 08 and saggar-fired. Colour from salt, copper carbonate, and steel wool. He says 'The wonderful gray color comes from a fairly tight saggar fit to the pot, and develops in the cooling phase after firing, but still remains mysterious and elusive to me.'

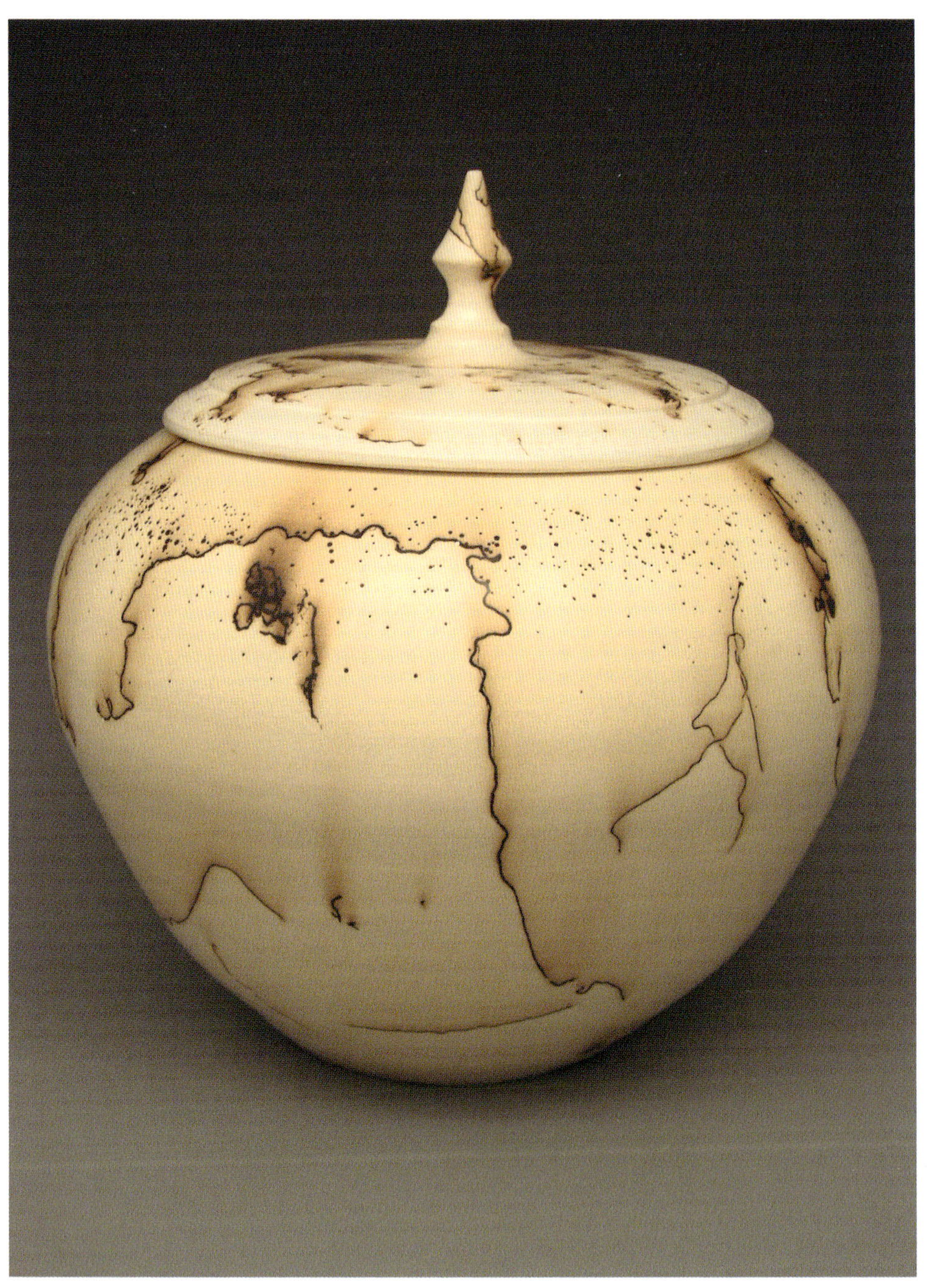

Horsehair Urn by Sumi von Dassow, 2006. Ht: 35 cm (14 in.), dia: 30 cm (12 in.). Wheel-thrown paper-clay with brushed terra sigillata, burnished with a chamois-leather. Bisque-fired to cone 010; reheated in an electric kiln to cone 018; removed from kiln with kevlar kiln gloves, horse-tail hair applied, then sprinkled with sugar.

Chapter 7

Raku firing techniques: horsehair and naked raku

Horsehair firing

Horsehair firing is fun, exciting, quick and distinctive. The process couldn't be simpler – take a hot pot out of the kiln and place hairs on it. As the hairs burn they shrivel up and twist around to leave squiggly black marks on the pot. The hairs don't actually have to be from a horse, though the hairs from a horse's tail are long and coarse and make nice, long, heavy marks. However, any kind of hair or fur can be used, including human hair, dog or cat hair or fur, or even feathers.

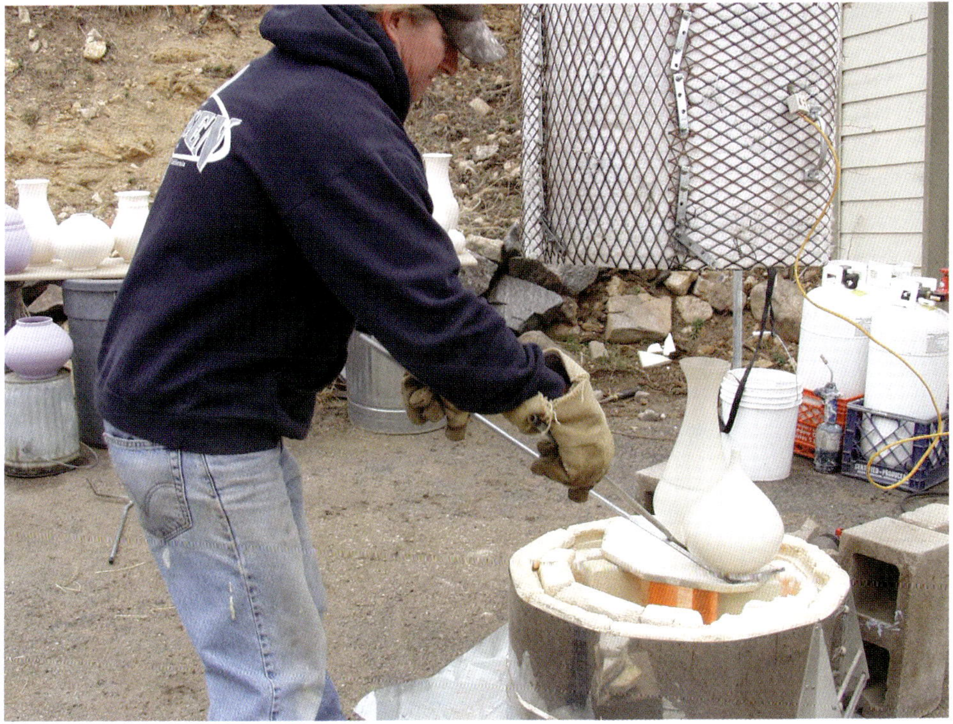

Travis Thomason removes a hot pot from the kiln carefully to avoid scratching the burnish. It will be placed on a piece of soft brick, which causes less thermal shock to a hot pot than a cold, hard brick.

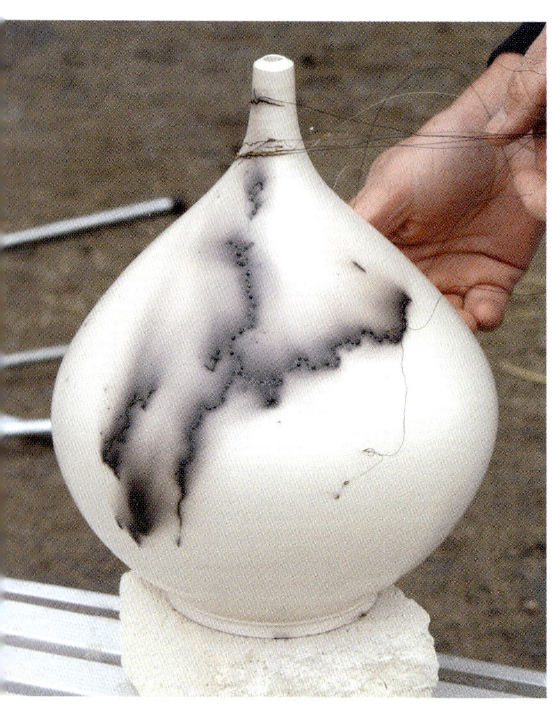

Horsehair, either individual hairs or in small bundles, is quickly applied to the hot pot.

Travis Thomason throws his large pots out of smooth white stoneware (B-Mix from Laguna Clay company), paints them with three coats of terra sigillata made from OM-4 ball clay, and bisque-fires to cone 010. He loads three large pots and one small one into a raku kiln, then heats them to 590–620°C (1100–1150°F). Because the firing is done outdoors in varying conditions, he's never sure exactly what temperature will work perfectly – so he removes the small pot first to test on, and adjusts the temperature up or down if needed. To keep a hot pot from cracking when placed on a cold surface he places it on a piece of soft insulating brick – another option is to actually preheat a piece of kiln shelf to place each pot on. He wraps handfuls of horsehair around the hot pot, sometimes blowing the smoke away so the pot doesn't get too black, until he is satisfied with the pattern. If the temperature is too high, the burning hairs produce a lot of smoke which leaves grey smudges around the black horsehair lines. Some potters like this effect, but Travis prefers the horsehair lines more distinct, so he may lower the temperature of his kiln and wait longer before placing hairs on the next pot he removes. On the other hand, if the temperature is too low the hairs won't burn, or they will burn without leaving a distinct mark.

While Travis fires his horsehair pots using a raku kiln, it is also possible to use an electric kiln for this type of firing. I usually heat up my pots in a small (46 cm/18 in.) electric kiln for this process, because it is convenient. I set the kiln to cone 018, which is almost 700°C (1300°F). The first pot I remove is usually too hot to decorate right away – the hairs almost bounce off and refuse to burn until the pot has cooled a bit – but this way the kiln can be turned off before the first pot is removed, and though slowly cooling it will stay hot enough to decorate several pots. For safety's sake it is important to turn an electric kiln off before reaching into it with metal tongs! Though the use of an electric kiln makes it very simple to decorate small pots with horsehair, it is a little tricky to do large pots this way. Most electric kilns are top-loading, so to remove a pot you must stand over it and reach in while heat radiates up at you. If your pots are small, you can place them on a shelf towards the top of the kiln, so that you can reach the pot with tongs without leaning over the kiln. But a large pot

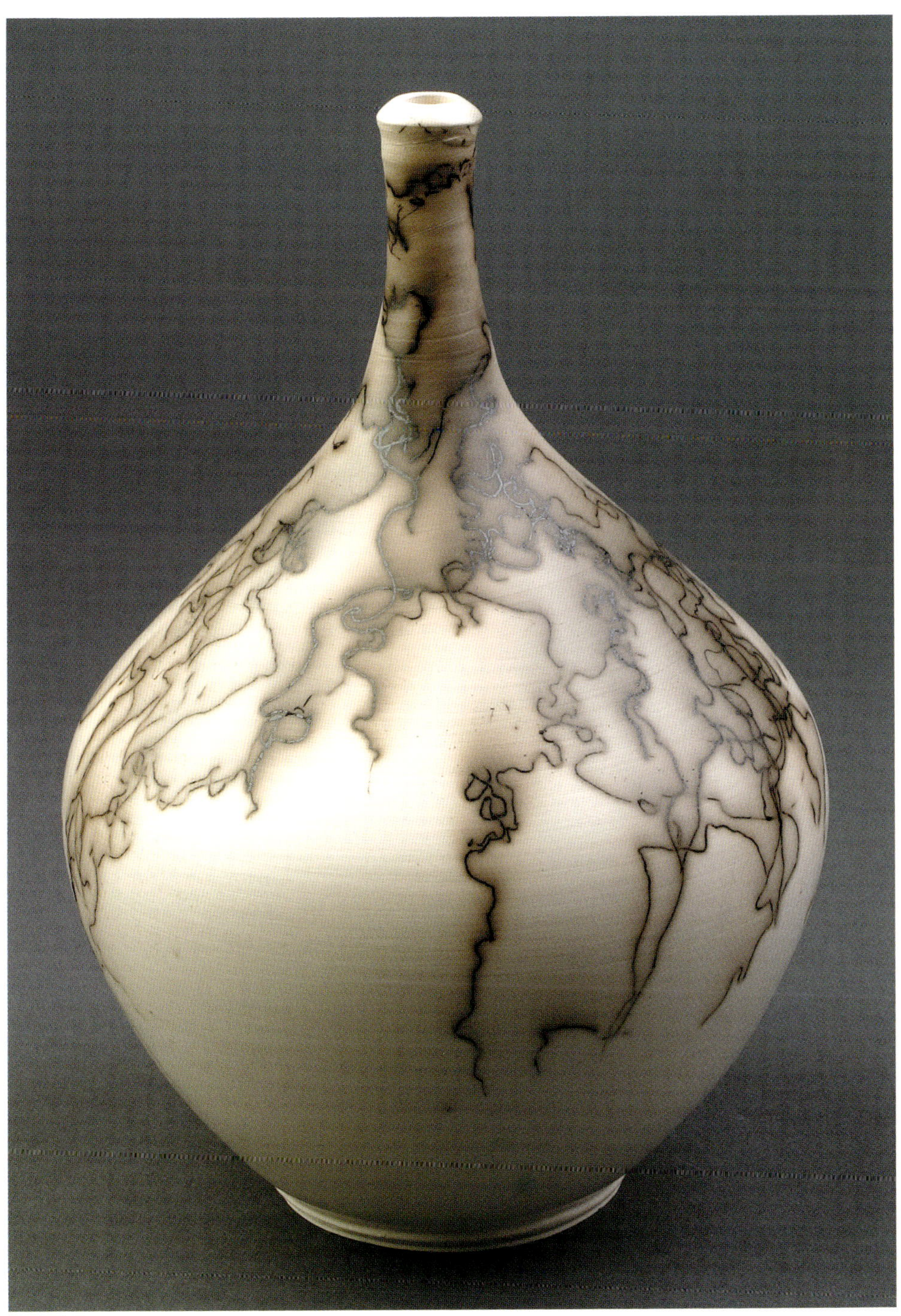

Tear-shaped Bottle by Travis Thomason. Ht: 25 cm (10 in.), dia: 20 cm (8 in.). Horsehair-fired vessel. Thrown B-Mix clay with brushed terra sigillata. *Photo by Ken Juliano.*

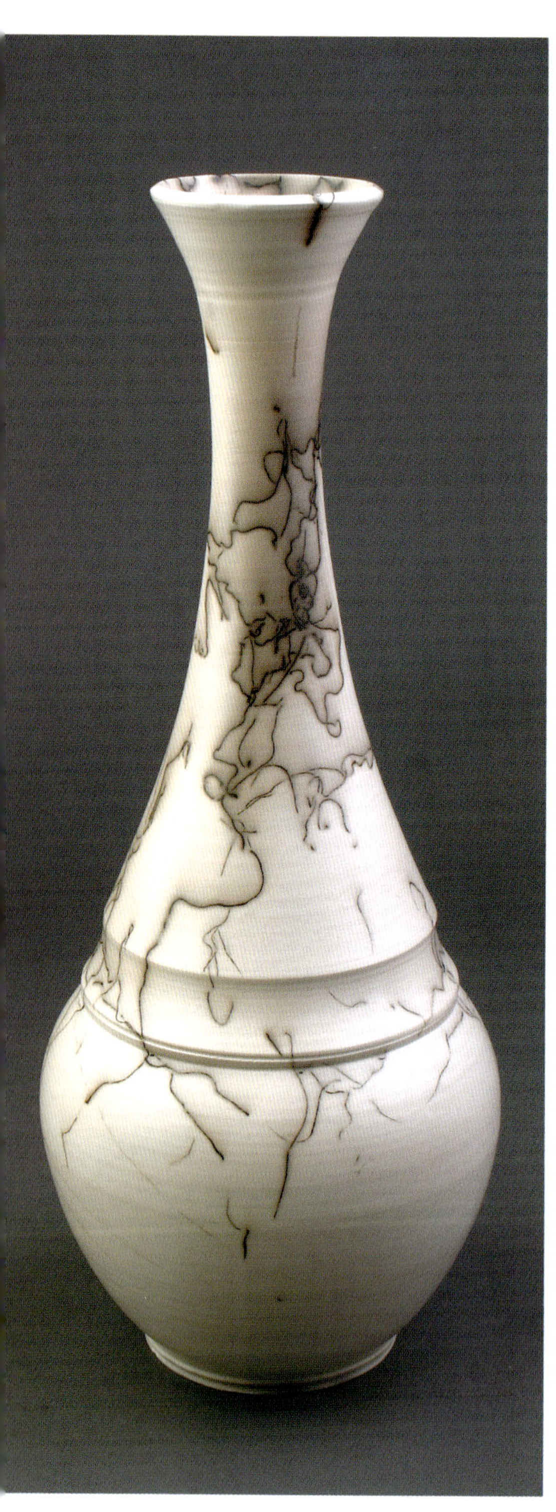

requires reaching the tongs straight down into the kiln, then lifting the pot high enough to clear the kiln wall. For the same reason it is inadvisable to use a large electric kiln for this type of firing – the heat radiating from an open 60 cm (24 in.) kiln at 700°C (1300° F) is considerably greater than that from a 46 cm (18 in.) kiln. If you are considering using an electric kiln for horsehair firing – or for any kind of raku – you should keep in mind that it may shorten the life of the bricks lining the kiln to open it while it is hot.

The biggest challenge to horsehair firing may be finding a source of horsehair. If you live in a rural area, visit any stable and you ought to be able to find a horse owner willing to part with some tail hair – especially if you offer a decorated pot in exchange. If you live in the city, you might not have such ready access to this commodity. You could seek out rental stables, search second-hand stores for old furniture

LEFT *Tall Vase* by Travis Thomason. Ht: 43 cm (17 in.), dia:18 cm (7 in.). Thrown B-Mix clay with brushed terra sigillata. *Photo by Ken Juliano.*

RIGHT, TOP *Golden Horsehair Vessel* by Charles and Linda Riggs, 2006. Ht: 15 cm (6 in.), w:18 cm (7 in.). White stoneware painted with white terra sigillata and polished with soft cloth; bisque-fired to cone 010; removed from kiln at 2012°C (1100°F); horse-tail hair applied; sprayed with ferric chloride solution.

RIGHT, BOTTOM Horsehair vase by Edgeworth Barnes. W: 13 cm (5 in.), ht: 12 cm (4½ in.). Wheel-thrown, burnished and fired with horsehair, feathers and sugar.

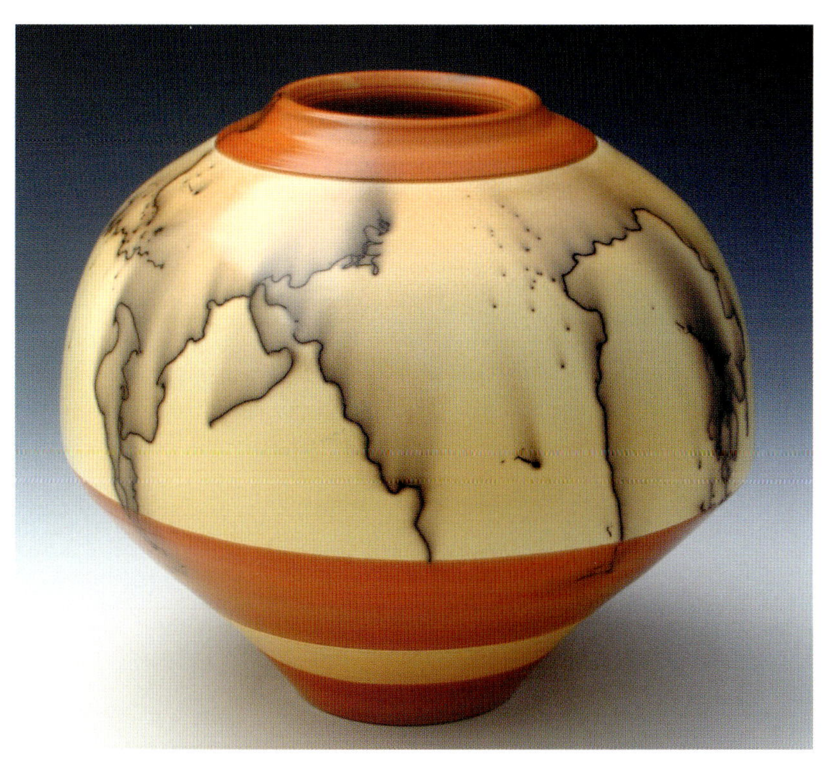

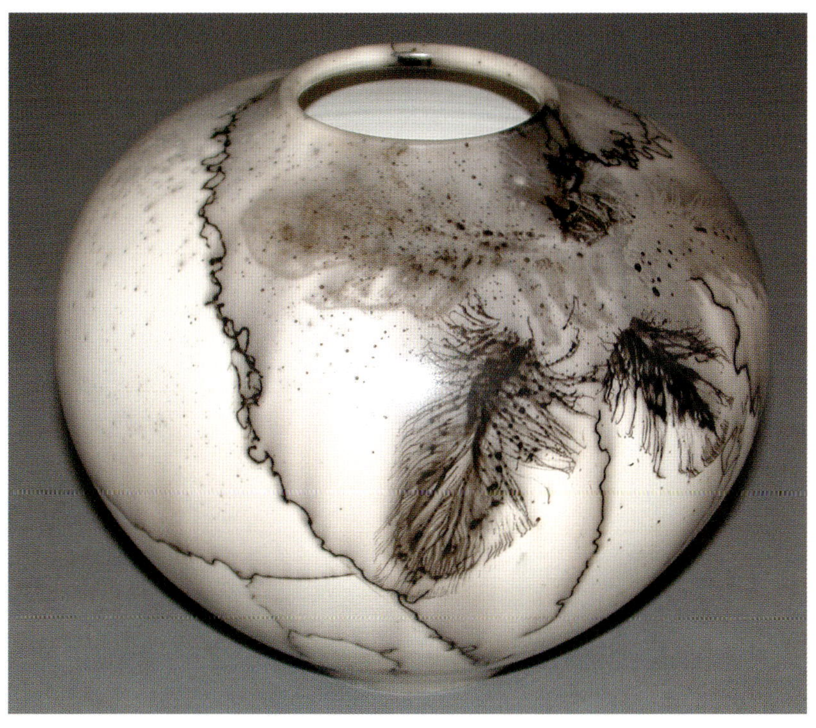

stuffed with horsehair, or ask a violin repair shop to save used horsehair when restringing bows. However, a quick internet search will probably yield a nearby mail-order source for horsehair, at least in the USA where horsehair is used for many crafts. If you simply cannot find horsehair, long human hair works much the same way, but leaves a finer mark. Dog or cat hair, even from a long-haired breed, is much shorter and leaves correspondingly short marks.

Feathers are a lot of fun to decorate with and leave a lovely mark, but they take a little more careful planning. If an entire feather is placed on a hot pot and left to burn it will create a large black tarry smudge. Instead, the feather must be drawn along the pot like a brush, or touched down and lifted up quickly. One last material to try is sugar – sprinkle a little on after you are finished with the hair or feathers, and you'll get a charming speckled effect. It will smell much nicer, too.

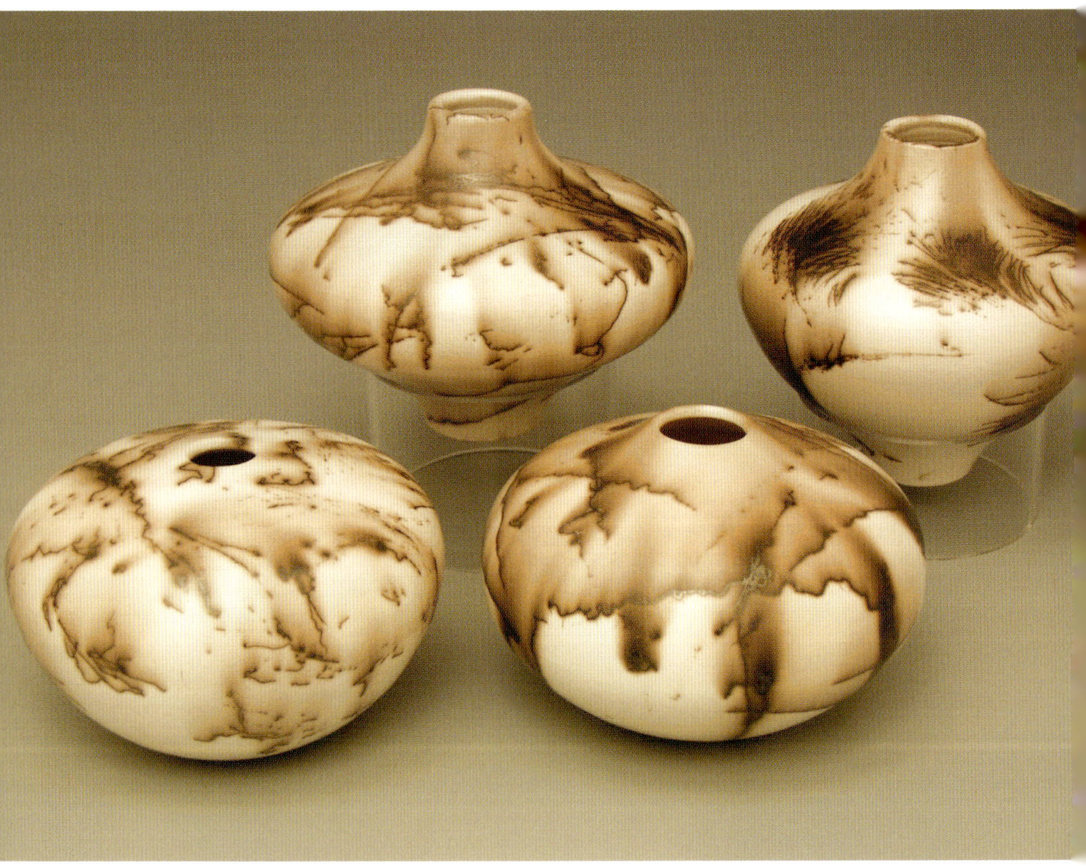

Four horsehair pots by Sumi von Dassow. Dia: approx 15 cm (6 in.). Wheel-thrown paper-clay with brushed terra sigillata; bisque-fired to cone 010; heated in an electric kiln to cone 018, with various types of hair applied. Clockwise from upper right: feathers, horsehair, dog hair and human hair.

Naked raku

Naked raku refers to pots which are fired much like traditional raku – removed hot from the kiln and placed in a reduction barrel – but the end product has no glaze, though it is marked with the distinctive raku crackle pattern. This is accomplished by coating the pot with slip, with or without a covering coat of glaze. After the firing, when the pot is removed from the reduction barrel, this coating falls off or is scraped off, leaving behind a crackle pattern on the bare clay.

For her naked raku pots Jan Lee uses an ungrogged white stoneware (Laguna's B-Mix cone 5), which she smooths with a rib and then paints with a thin coat of terra sigillata. She bisque-fires to cone 08, then pours on one or two layers of 'Wally's Slip', followed when dry enough to touch by a thin layer of clear raku glaze. After the pots have dried overnight they are preheated in an electric kiln, then transferred to a raku kiln and slowly brought to 737°C (1350°F). Lee watches the pots as they heat and when she sees an 'orange peel effect'– a slight pitting – in the glaze surface she removes the pot to a metal bin with lots of wood shavings in the bottom, adds more shavings on top, and covers the bin. As the pot slowly cools in the reduction chamber the resist slip and glaze coating shrink and crack, allowing smoke to penetrate the cracks and drawing carbon deep into the underlying clay body. As the pot cools, the slip and glaze coating is peeled away leaving behind a beautiful crackle pattern on the unglazed surface. Sometimes the coating sticks a bit in spots and has to be scraped or scrubbed carefully away.

For interesting effects, you can try coating the pot with slip, then applying glaze to only part of the pot. Both Jan Lee and Wally Asselberghs of Belgium use this technique, applying a thin coat of slip and splashing glaze over it. In the reduction process, the smoke penetrates the thin layer of slip and leaves the pot mostly black with white only where the glaze was splashed over the slip. Another interesting technique is to use sgraffito to draw through the glaze covering the slip layer, creating a deliberate pattern instead of the usual random pattern of crackle lines.

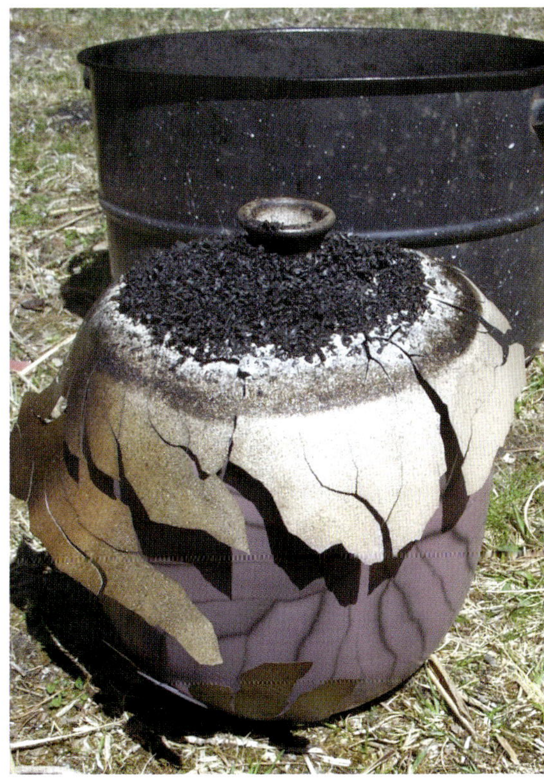

RIGHT A naked raku pot by Jan Lee, just removed from its reduction barrel. The coating of slip and glaze is already peeling off as the pot cools.

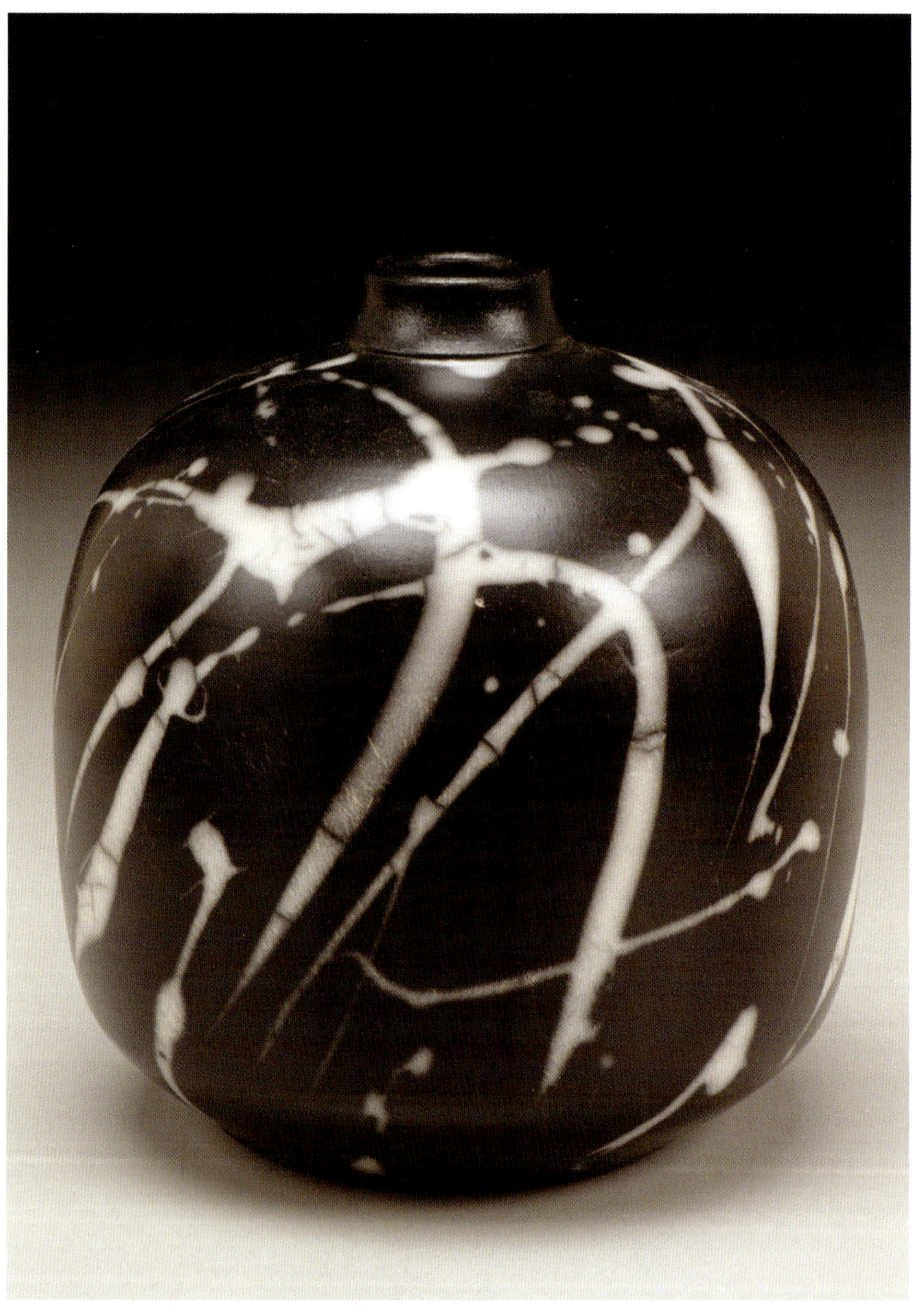

Night by Jan Lee, 2005. Ht: 18 cm (7 in.), dia: 13 cm (5 in.). This naked raku pot had only one coat of slip poured over it, then glaze was splashed on with a loaded brush before it was fired. *Photo by Tim Barnwell.*

Recipes

Wally's Slip

Highwater Raku Clay	50
EPK Kaolin	30
Silica	20

This is Wally Asselberghs' recipe for American potters. In Europe he uses a clay called Terre de la Puisaye, mined in France in the area of St Amand.

Robert's Resist Slip

EPK Kaolin	60
Silica	40

Jan Lee uses this recipe also, and she thinks it comes off cleaner. She suggests two quick coats on pots bisque-fired to cone 08, while usually one coat is enough on 010 bisque.

Raku Overglaze

Gerstley Borate	35
Ferro Frit 3110	65

Some potters do not use a covering glaze coat for their naked raku pots. Without a covering glaze, the slip falls off the pot very easily after the firing. This is both a blessing and a curse. On the one hand, there is no chance that glaze will seep through the slip and stick to the pot, ruining it or making for arduous work removing it. On the other hand, it can be very tricky to get the slip to stick long enough to get the pot into the kiln, then out of the kiln and into the reduction barrel. For this method a different slip recipe is recommended. It can be modified to

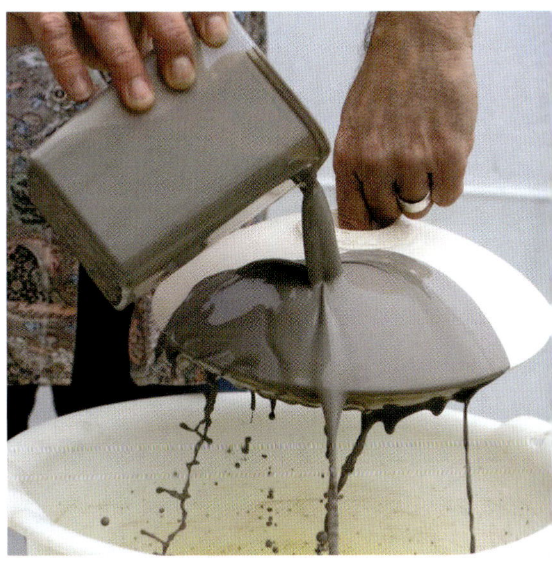

Wally Asselberghs pours slip on to a pot for naked raku firing. He will next pour on a coat of thin glaze. The rim is masked with tape which will be peeled off before firing, to leave the rim black after firing. *Photo by Heleen Callens.*

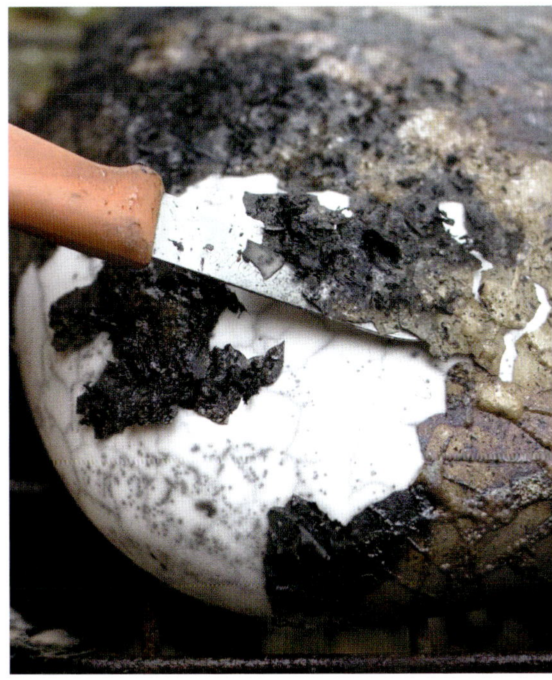

RIGHT Wally Asselberghs peels the 'shell' of slip, glaze and splashed glaze from a fired naked raku pot. *Photo by Heleen Callens.*

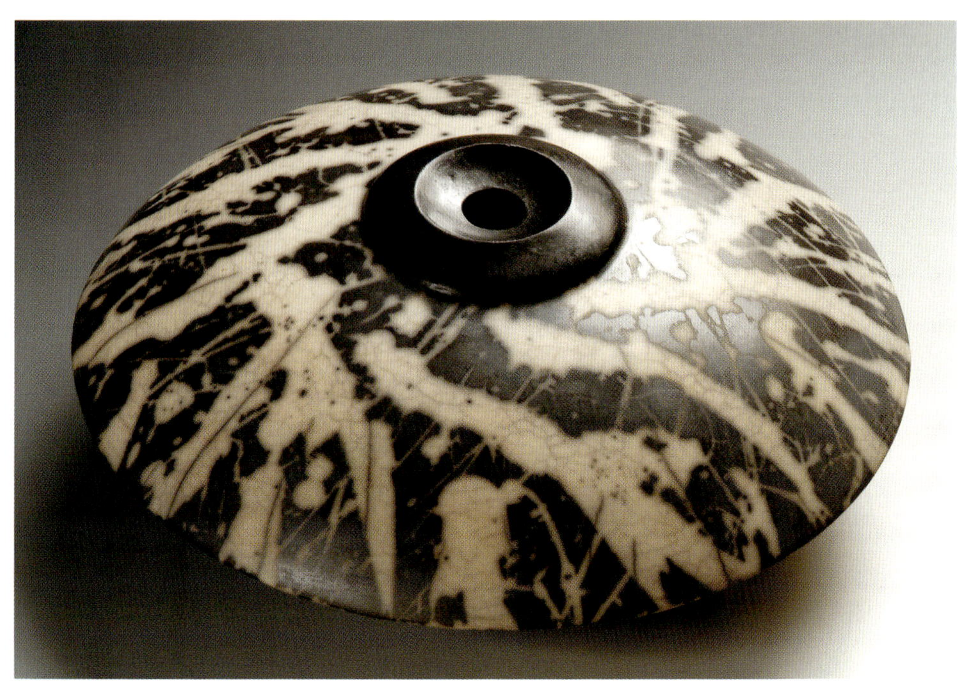

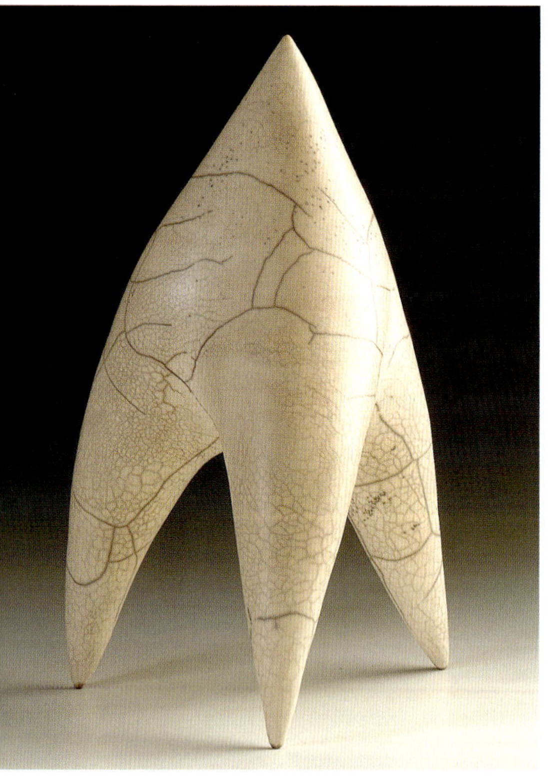

ABOVE *UCO* (*Unidentified Ceramic Object*) by Wally Asselberghs, 2005. Ht: 11 cm (4½ in.), dia: 27 cm (10¾ in.). The pot was covered with slip, then splashed with glaze before firing. *Photo by Lucille Feremans.*

LEFT *Monstruonzo* by Wally Asselberghs, 2004. 36 x 27 x 28 cm (14 x 10¾ x 11 in.). Naked raku. *Photo by Lucille Feremans.*

make it stick better by adding more alumina hydrate or a bit of CMC gum. You may find that it is difficult to get this slip to stick on to vertical surfaces, and even more so on the lower half of a rounded pot. Any place where the slip falls off will be black after it emerges from the reduction barrel. Shallow bowls or plates, or pots with a wide flat shoulder may be the most effective forms to try out this technique.

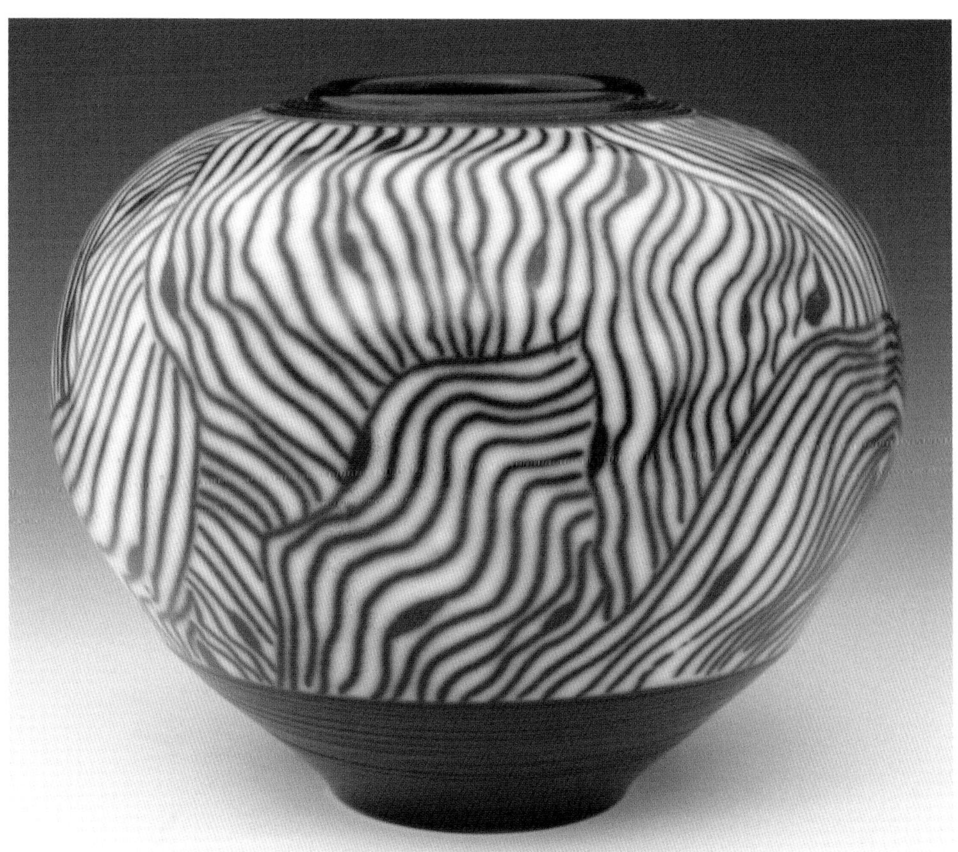

Naked Raku Vessel by Charles and Linda Riggs, 2003. Ht: 23 cm (9 in.), w: 25 cm (10 in.). Naked raku pop-off slip. White stoneware painted with white terra sigillata and polished with soft cloth; bisque-fired to 010. The top and bottom of the pot were protected by tape. It was then dipped in thick pop-off slip (with no glaze). The tape was removed and the pot raku-fired to 816°C (1500°F). Pulled and heavily reduced in newspaper. After cooling the slip pops off the pot leaving crackle patterns on the surface.

Riggs' Pop-off Slip

4.4 kg (9.7 lb) Hawthorne Fire Clay 35 mesh
1.44 kg (3 lb) EPK Kaolin
700 g (1.6 lb) Alumina Hydrate
5 litres (9 pints) water

To add CMC, first mix the CMC powder into hot water and let it dissolve. Add small amounts to a portion of the slip immediately before using – and keep notes. Be forewarned that this is a tricky technique to master, and all kinds of variables will affect it including humidity, air temperature, weather, clay body and probably minerals in the water.

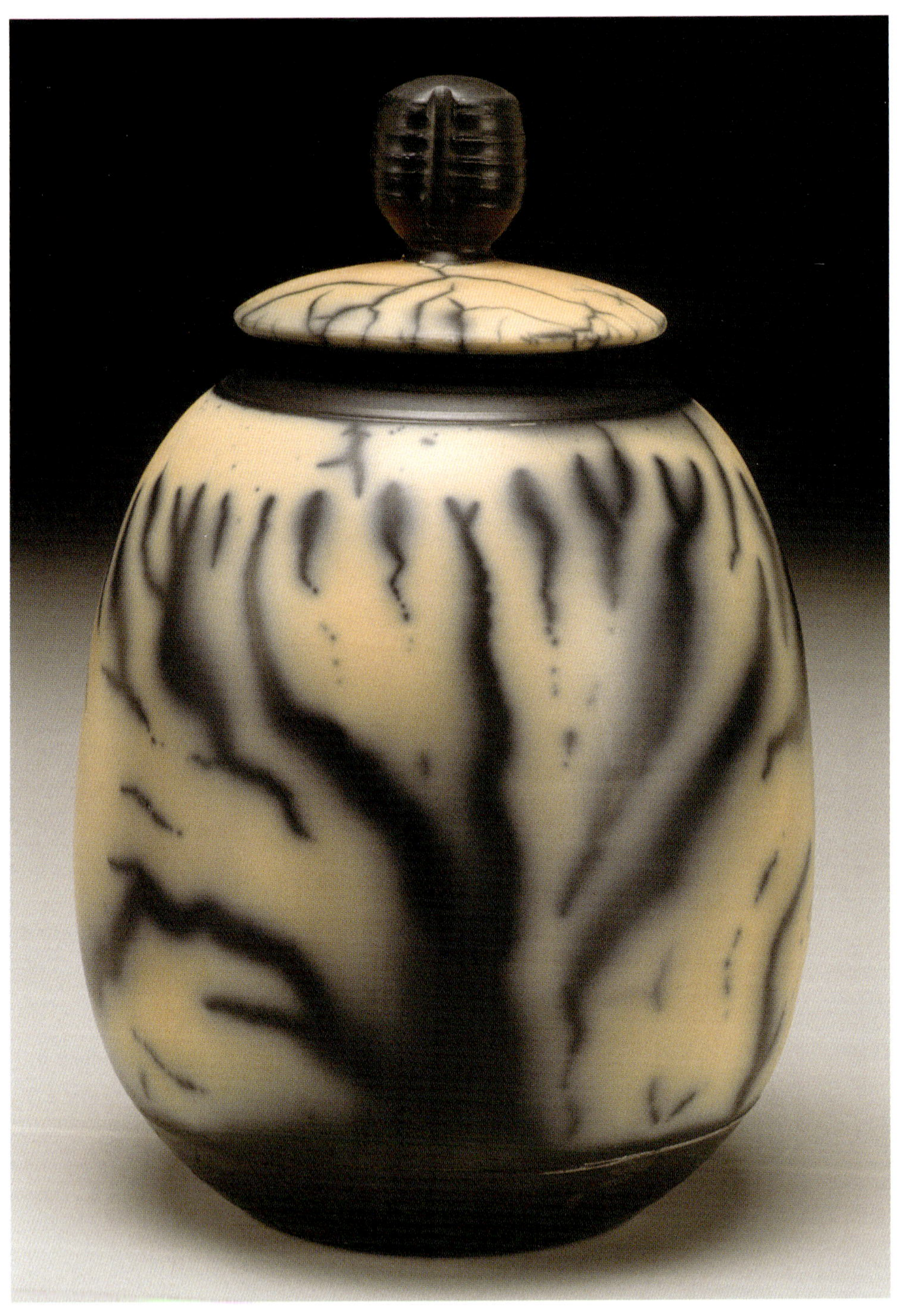

Lidded Form by Jan Lee, 2006. Ht: 33 cm (13 in.), w: 25 cm (10 in.). Naked raku coated with white terra sigillata coloured with 1 tsp. of crocus martis per cup. *Photo by Tim Barnwell.*

Chapter 8

Finishing touches

While a burnished pot is finished after firing, some potters like to enhance or protect the surface with some form of wax or acrylic coating. Some want to make their burnished pots even more shiny, while others simply want to protect the surface from being stained by dirty fingers or accidental marks. It is also possible that such sealers can preserve colours which may otherwise fade over time if exposed to direct sunlight. With my own firing methods I haven't observed any fading, but your experience with your particular firing method will indicate whether this is a concern. Paste wax, floor wax and tile sealer are the products most commonly used for this purpose.

Paste wax

There are numerous brands, available in solid or liquid form. One brand some potters prefer is Trewax. Solid paste wax needs to be applied with a cloth and then buffed. A drawback is that on a textured surface it is impossible to buff inside cracks and crevices, so it will leave whitish residue in these spots. This can be remedied by heating the pot in a warm oven. Liquid paste wax does not leave a residue. Any paste wax may dull over time and may need to be buffed again or reapplied. An advantage of using paste wax is that if you change your mind about coating the surface of your pot, the wax can be removed by heating in an oven to a temperature low enough that it won't affect the fired colours. The complementary disadvantage is that storage in a warm room may be enough to degrade the wax coating.

Floor wax

A brand to try is Pledge (formerly Future). This is not a wax but a clear acrylic product. Applied full strength it can look almost like a glaze, but diluted with water (as little as a tablespoon in half a cup of water) it can enhance the sheen, protect the surface from staining and allow the pot to be cleaned with water. It can be brushed on with a sponge brush, or wiped on with a cloth. Once applied, this coating is relatively permanent and cannot be removed without refiring the pot and changing the fired colours.

Tile sealer

Dal-Tile makes an acrylic sealer that is used to seal terracotta tile to protect it from the elements and for ease of cleaning. Since this product is intended for use on fired clay, it seems appropriate to use it on burnished pots. Like the floor wax, tile sealers can be applied with a sponge brush or a cloth. Dal-Tile's Penetrating Sealer doesn't change the surface at all, but soaks in and makes it

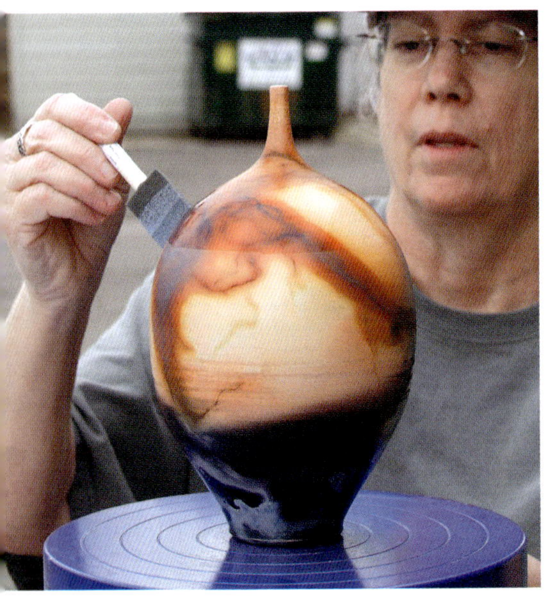

Linda Dadisman finishes her saggar-fired pots by applying a thin coat of tile sealer.

more watertight. Another product, called finishing sealer, comes in gloss or matt, and if applied heavily will make the surface of the pot appear glossier. This product can also be used to fill in hairline cracks by pouring it into the pot and swirling it around the entire inside before pouring out the excess. Similarly, if your terra sigillata develops surface cracks or partially peels off during firing, this product can be applied to help stabilise the surface.

Varnishes

Some potters like to pour polyurethane varnish inside their pots to make them watertight. Of all the products mentioned, varnish is the most effective sealer, but it changes the surface it is applied to more obviously than the other products.

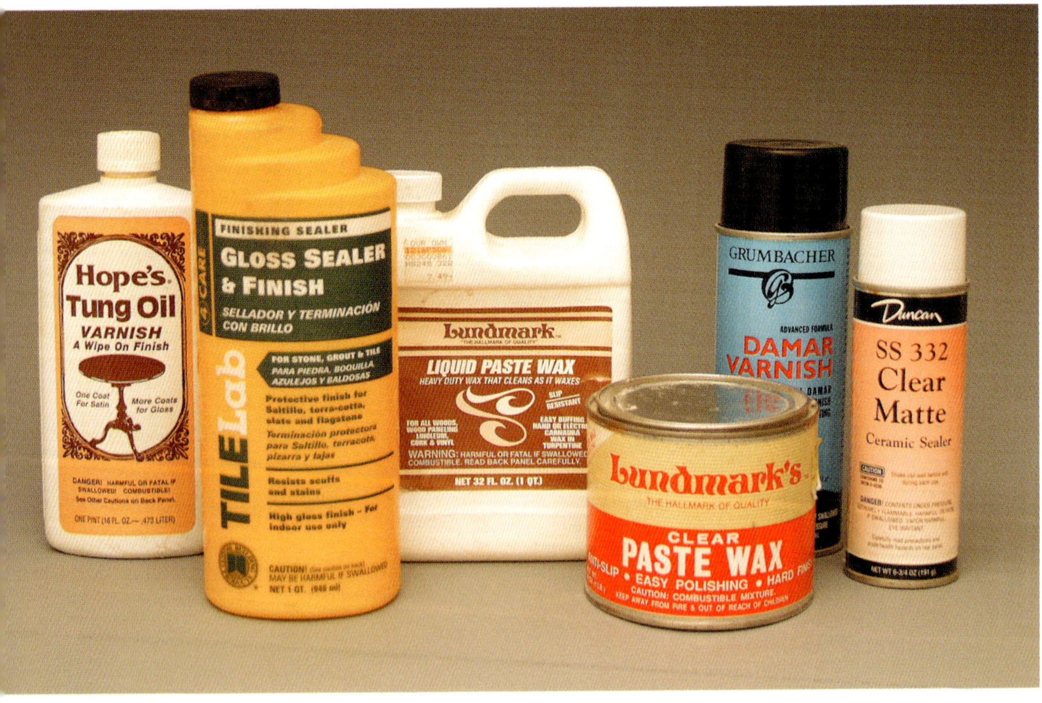

Various waxes and sealers can be used to protect or enhance the surface of a fired burnished pot.

Varnish may have a yellowish tint, so it cannot be recommended as a finish for the outside surface. You will have to decide whether the advantage of sealing the inside is worth the change to the appearance inside the pot.

Acrylic spray glaze

Some ceramic manufacturers make an acrylic spray, in gloss or matt, which can be applied to unglazed surfaces. A spray is easy to apply, but you do have to be careful not to spray too heavily or allow drips to form.

Paint

While paint is not a sealer, there are flaws which can be fixed with a careful application of paint. For instance, occasionally small patches of terra sigillata may peel off during firing, or bits of lime in the clay can cause pop-outs in the surface. Either acrylic or oil paint can be used to match the surrounding colour, though the burnished texture will not be matched. Many potters might consider it cheating to add paint; however, on a purely decorative piece of art, this issue is an individual judgment call.

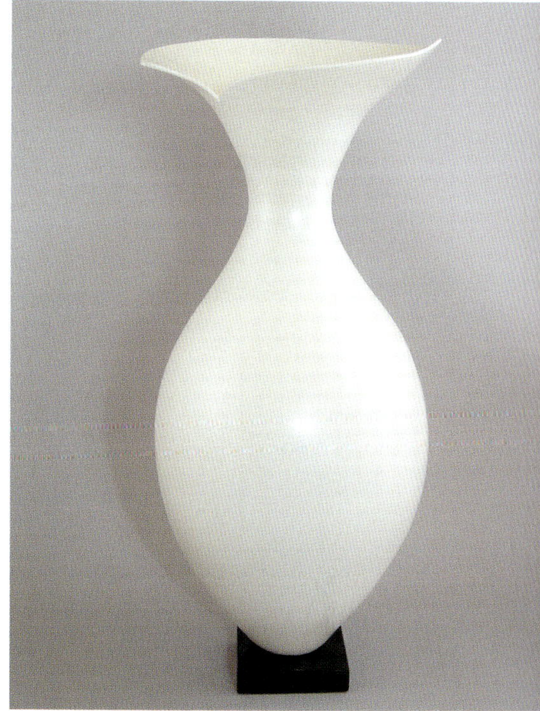

RIGHT, TOP *Egret* by Carla Kappa, 2004. Coil-built earthenware (CT-3 clay), burnished with pebble, fired to cone 04. Ht: 63.5 cm (25 in.), dia: 28 cm (11 in.).

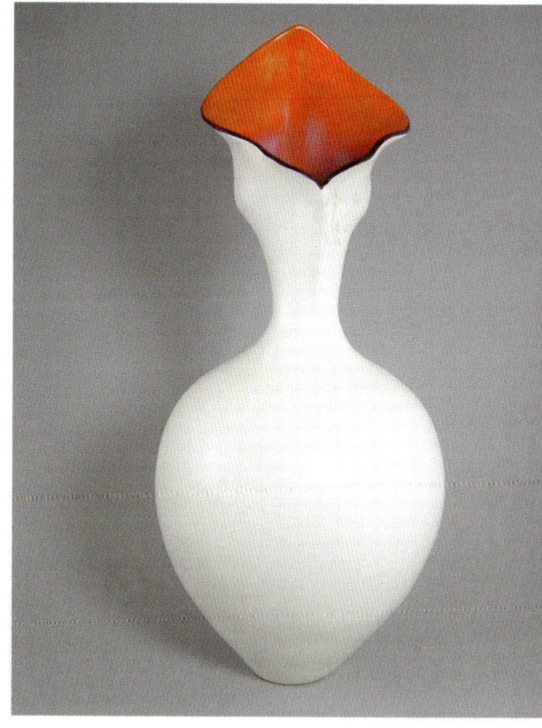

RIGHT *Poppy* by Carla Kappa, 2006 Ht: 66 cm (26 in.), dia: 35.5 cm (14 in.). Coil-built white earthenware (CT-3), burnished with pebble, fired to cone 04 with a 'cold finish' applied inside after the final firing. 'Cold finish' refers to any finish that isn't fired on, such as acrylic paint.

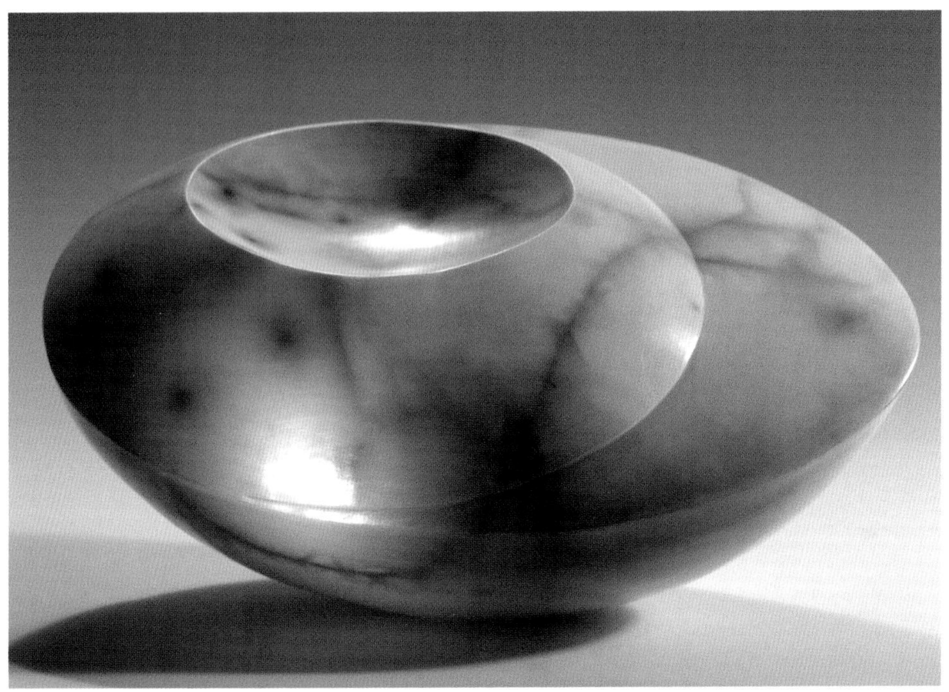

Marbled Kernel Form by Antonia Salmon, Ht: 16 cm (6¼ in.). Hand-built from white stoneware clay, burnished using teaspoon and plastic rods at least 4 times. Bisque-fired to 1000°C (1832°F) and then smoke fired several times. Polished with wax polish after firing.

Gold leaf

Unlike paint, this product can't be used to hide a flaw such as peeling terra sigillata, but it can be applied over an unsatisfactory area to disguise the flaw and embellish the surface of the pot. Gold leaf is easy to apply, using a special type of glue called 'size'. Genuine gold leaf is of course quite expensive, but gold-coloured leaf is readily available and inexpensive. Gold leaf has been used to cover repairs in valuable pieces of pottery for centuries, and it can certainly be used to enhance work even where no repairs are needed.

One thing to keep in mind when deciding whether to apply any protective sealer to your burnished pots is that, while protecting the surface of your pot from accidental stains, waxes and sealers may alter some colours. For instance, pit-fired pots which emerge from the firing with a metallic black colour may lose the metallic quality or become blotchy when waxed. In addition, of course, the application of any product changes the character of the surface and may potentially make it less valuable to collectors. Try various sealers on broken or less favoured pots before putting anything on your masterpiece.

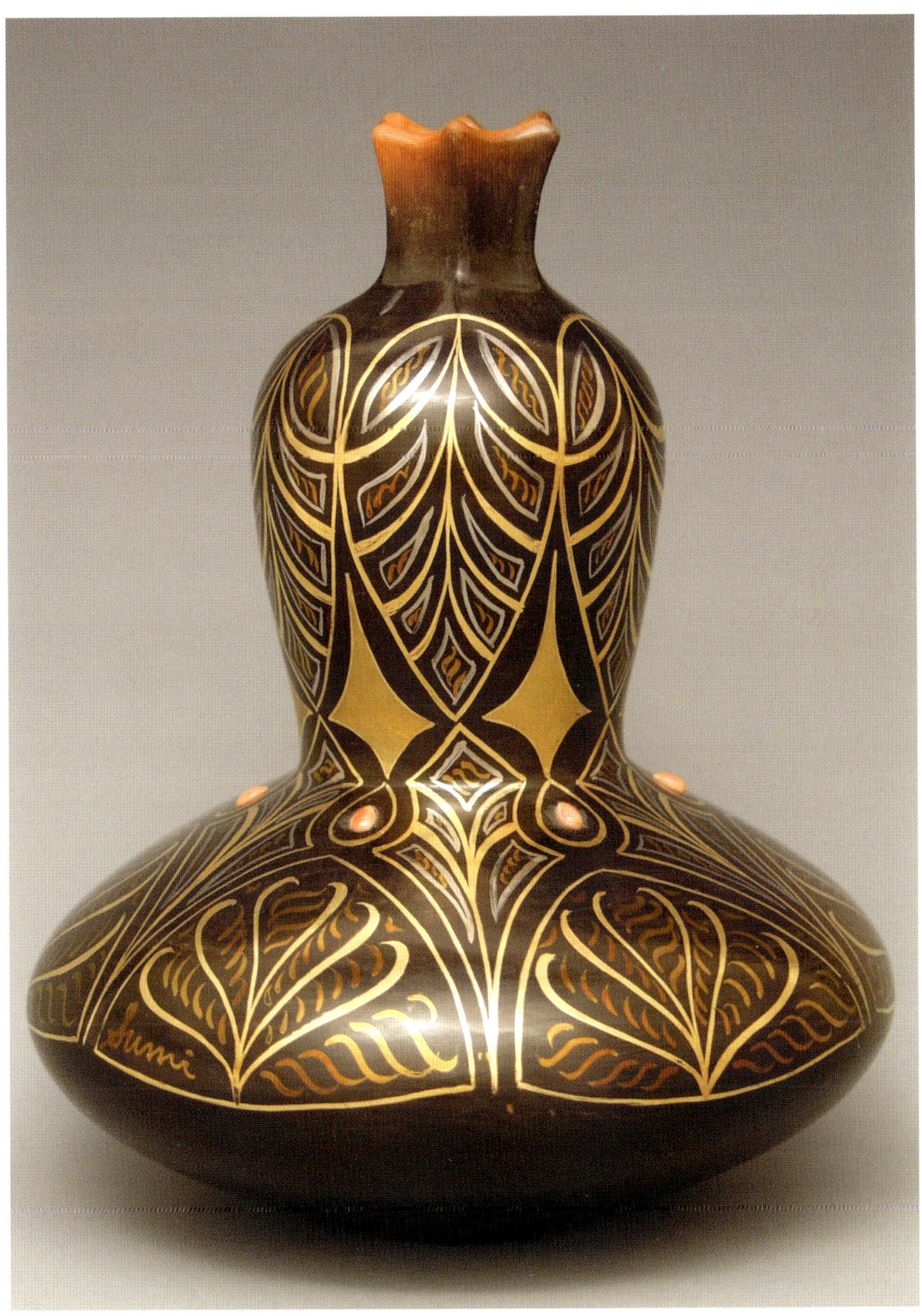

Coral Starflower, by Sumi von Dassow. 24 × 19 × 19 cm (9½ × 7½ × 7½ in.). Coil-built from red stoneware (Navajo Wheel Clay from Industrial Minerals Co.), burnished with a stone, painted with lustres (gold, platinum, and bronze), and black-fired. Gold leaf and pieces of coral applied after firing.

Bibliography

Barley, Nigel, *Smashing Pots: Works of Clay from Africa* (Smithsonian Institution Press, Washington, DC, 1994)

Blandino, Betty, *Coiled Pottery, Traditional and Contemporary Ways* (Krause Publications, Iola WI, 2003)

Boardman, John, *The History of Greek Vases* (Thames and Hudson, New York, NY, 2001)

Branfman, Steven, *Raku: A Practical Approach* (Krause Publications, Iola WS, 2001)

European Ceramic Work Centre, Edited by Anton Reijnders, *The Ceramic Process: A Manual and Source of Inspiration for Ceramic Art and Design* (A & C Black, London, 2005)

Freestone, Ian and Gaimster, David, *Pottery In the Making: Ceramic Traditions* (Smithsonian Institution Press, Washington, DC, 1997)

Hessenberg, Karin, *Sawdust Firing* (University of Pennsylvania Press, Philadelphia, 1994.)

Higgins, Reynold, *Minoan and Mycenaean Art* (Thames and Hudson, New York, NY, 1985)

Medley, Margaret, *The Chinese Potter: A Practical History of Chinese Ceramics* (Phaidon Press, Ltd., London, 1989)

Noble, Joseph Veach, *The Techniques of Painted Attic Pottery* (Watson-Guptill Publications, New York, NY, 1965)

Peckham, Stewart, *From This Earth: The Ancient Art of Pueblo Pottery* (Museum of New Mexico Press, Santa Fe, NM 1990)

Perryman, Jane, *Naked Clay: Ceramics Without Glaze* (A & C Black, London, 2004)

Perryman, Jane, *Smoke Firing: Contemporary Artists and Approaches* (A & C Black, London/ University of Pennsylvania Press, Philadelphia, 2008)

Peterson, Susan, *Pottery by American Indian Women: The Legacy of Generations* (Abbeville Press, New York, NY, 1997)

Reina, Ruben E. and Hill, Robert M., *The Traditional Pottery of Guatemala* (University of Texas Press, Austin and London, 1978)

Rice, Prudence M., editor *The Prehistory and History of Ceramic Kilns* (The American Ceramic Society, Westerville, OH, 1997)

Schreiber, Toby, *Athenian Vase Construction: A Potter's Analysis* (J. Paul Getty Museum, Santa Monica, CA, 1999)

Schofield, J.F., *Primitive Pottery: An Introduction to South African Ceramics, Prehistoric and Protohistoric* (The South African Archaeological Society, Cape Town, South Africa, 1947)

Turner, Anderson, *Raku, Pit and Barrel, A Ceramic Arts Handbook* (The American Ceramic Society, Westerville, OH 2008)

von Dassow, Sumi, *Barrel, Pit and Saggar Firing: A Ceramics Monthly Handbook* (The American Ceramic Society, Westerville, OH 2001)

Watkins, James and Wandless, Andrew, *Alternative Kilns and Firing Techniques* (Lark Books, New York, 2004)

List of suppliers

USA & Canadian Suppliers

Laguna Clay Company
14400 Lomitas Ave
City of Industry, CA 91746, USA
Tel: (626) 330-0631
Freephone: (800) 4-LAGUNA
Ohio location at: (800) 762-4354
www.lagunaclay.com
(Many local distributors throughout the United States.)
B-Mix (white stoneware), Miller 3.

Industrial Minerals Company
7268 Frasinetti Road
Sacramento, CA 95828, USA
(916) 383-2811
Navajo Wheel Clay.

Mile-Hi Ceramics
77 Lipan
Denver, CO 80223, USA
(303) 825-4570
www.milehiceramics.com
CT3 clay, Ash white, Biz Bod clay.

Highwater Clay
Highwater Clays, Inc.
600 Riverside Drive
Asheville, NC 28801, USA
(828) 252-6033
www.highwaterclays.com
Highwater Raku Clay and Orangestone clay.

Axner Ceramic Supply
490 Kane Ct.
Oviedo, FL 32765, USA
407-365-2600/800-843-7057
www.axner.com
Axner Super Kilns.

American Art Clay Co.
6060 Guion Road,
Indianapolis IN 46254, USA
Tel: (317) 244 6871
www.amaco.com

Tuckers Pottery Supplies Inc.
15 West Pearce St., Unit 7
Richmond Hill
Ontario, Canada L4B 1H6
Tel: 800 304 6185
www.tuckerspottery.com

UK Suppliers

Ceramatech Ltd.
Units 16 & 17 Frontier Works
33 Queen St, Tottenham North,
London N17 8JA, UK
Tel: 0208-885-4492
www.ceramatech.co.uk
T-Material, White St. Thomas Clay, CT1103 Porcelain etc.

Potclays Co. Limited
Brickkiln Lane, Etruria
Stoke-on-Trent, Staffordshire,
ST4 7BP, UK
Tel: 01782 219816
Fax: 01782 286506
www.potclays.co.uk
T-Material, White St. Thomas Clay, CT1103 Porcelain etc.

Bath Potters' Supplies
18 Fourth Avenue, Westfield Trading
Estate, Radstock, Bath, BA3 4XE
Tel: 01761 411077
Fax: 01761 414115
www.bathpotters.co.uk

Scarva Pottery Supplies
Unit 20, Scarva Road Industrial Estate
Banbridge, Co. Down BT32 3QD
Tel: 018206 69699
Fax: 018206 69700
www.scarvapottery.com

Valentines Clay Products
The Sliphouse
18-20 Chell Street, Hanley
Stoke-on-Trent ST1 6BA
Tel: 01782 271200
Fax: 01782 280008
www.valentineclays.co.uk

Worldwide Suppliers

Keramikos
Oudeweg 153
2031 CC Haarlem
The Netherlands
Tel: 023 542 44 16
www.keramikos.nl

Colpaert Ceramic & Sculpture Materials
Groendreef 51
9000 Gent
Belgium
Tel: 09/226.28.26
www.colpaertonline.be

Cape Pottery Supplies
Unit A2, 11 Celie Road
Retreat Industria
Western Cape, 7945
South Africa
Tel: +27 21 701 1320/1
www.capepotterysupplies.co.za

Pottery Supplies
51 Castlemaine St.
Milton QLD 4064
Brisbane, Australia
Tel: +617 3368 2877
www.potterysupplies.com.au

Walker Ceramics
2/21 Resaerch Drive
Croydon South, Victoria 3136,
Australia
Tel: +61 (3) 8761 6322
www.walkerceramics.com.au

Contributing artists

Wally Asselberghs www.wallyasselberghs.be
Edgworth Barnes Email: potsedge@earthlink.net
Jane Burton www.burtonceramics.com
Joan Carcia www.jcarcia.com
Linda Dadisman Email: linda@milehiceramics.com
Ian Garrett Email: iangarrettceramics@gmail.com
David Greenbaum www.greenbaumpottery.com
Ashraf Hanna www.studiopottery.co.uk
Bill Hensler Email: hensler@velocitynetdsl.com
Carla Kappa kappa@carlakappa.com
Gabriele Koch www.ceramics-aberystwyth.com/gabriele-koch.html
Jan Lee www.southernhighlandguild.org/janpottery/
Ricky Maldonado www.rickymaldonado.com
Judith Motzkin www.motzkin.com
Carol Molly Prier www.PointReyesArt.com
Charles and Linda Riggs www.cclay.com/criggs/
Ron Rivera www.potpaz.org
Duncan Ross www.duncanrossceramics.co.uk
Antonia Salmon www.antoniasalmon-ceramics. co.uk
Bob Smith www.bobsmithraku.com
Travis Thomason www.travisthomasonpottery.com
Michael Wisner www.southwestpottery.com
Nancy Youngblood www.nancyyoungbloodinc.com

Glossary

Ball Clay – a type of highly plastic clay that is named after the habit of English miners of forming it into large balls for transportation

Bentonite – a form of clay characterized by being extremely absorbent

Bisque-firing – firing to a temperature below the maturing temperature of a clay body, usually a preliminary to glazing a pot

Black-hard – between leatherhard and bone-dry, when a pot is quite hard, almost dry, but hasn't yet lightened in color

Bone-dry – used to describe a pot that has completely air-dried, signified by the clay lightening in color

Broom corn – broom straw, available from craft suppliers for making handmade brooms

Burnished – polished

Calgon – a water softener, formerly often used in making terra sigillata, now formulated without phosphate and no longer usable for terra sigillata

Clay – fine-grained soil consisting of flat particles which trap water. Clay is plastic and workable when wet, able to hold a form when dry and hardens when fired. 'Clay' refers to the naturally occurring material which is mined and used in the production of a 'clay body'

Clay body – a mixture of naturally occurring clays and other materials, this is the proper term for what you buy in a bag from a supplier

CMC – carboxymethyl cellulose, a gum used as a binder in slips and glazes

Chamois – fine soft leather, useful for buffing terra sigillata

Cone – small pyramids or bars formulated from ceramic materials to melt at specific temperatures, depending on firing speed. Cones range from cone 021 to cone 12. It is more accurate to refer to firing temperatures in terms of cones than degrees, as the melting point of a cone will vary with firing speed. Cones can be used to automatically shut off an electric kiln, or they can be positioned so the potter can monitor them visually during a firing

Copper carbonate – a non-soluble source of copper used in saggar- and pit-firing, available from any ceramic supplier as a glaze colourant

Copper sulphate – a soluble source of copper for saggar- or pit-firing, sold by ceramic suppliers, or by garden suppliers for fertilizer

Darvan – Darvan 7 or Darvan 811 are deflocculants

Deflocculant – a chemical which weakens the electrical attraction between particles of clay, thus breaking up small clumps of clay and allowing the individual particles to float freely. Deflocculent is used to make casting slip and terra sigillata

Earthenware – low-fire clay, maturing around cone 06 to 04 (around 982°C/ 1800°F.), made by adding non-clay materials to natural clays to lower the maturing point. Talc is one of these non-clay materials

EPK – Edgar Plastic Kaolin, the standard kaolin used in the USA, but any kaolin should work provided it is tested first

Ferric chloride – a soluble source of iron oxide often used in saggar- or pit-firing, or sprayed onto hot raku or horsehair pots to create a golden colour. Sold liquid at electronic supply stores as etchant solution, or as granules to be mixed with water from scientific stores. Extremely toxic and caustic

Ferric sulphate – a soluble source of iron oxide which can be used in saggar- or pit-firing. Sold as fertilizer in garden supply stores. Not as toxic as ferric chloride

Giffin Grip – a piece of equipment used to centre clay for trimming, or for burnishing on the wheel

Greenware – unfired pottery

Glaze – a glass coating on the surface of fired pottery. Not used on burnished pottery

Grog – ground bisque-fired clay, added to clay bodies to improve workability, prevent large work from cracking while drying, and improve resistance to thermal shock in fast firing processes such as raku

Hydrometer – a device used to measure specific gravity in liquids

ITC coating – a material made by International Technical Ceramics which is used to increase the longevity of kiln walls, furniture and elements

Kaolin – a form of natural clay which is very pure and white

Kaowool – ceramic fibre, sold as a blanket, similar to fibreglass

Kevlar – fire-resistant material

Kiva – an underground or partly underground chamber in a Pueblo village, used by the men especially for ceremonies or councils. It is always entered from above by a ladder and it is a very important part of the life of Pueblo people. The step motif is very important in Pueblo pottery

Leatherhard – a state of clay between moist and dry, when the clay is firm but not dry. It is easy to cut, scrape or carve leatherhard clay, but it can't be moulded

Lye – sodium hydroxide, a corrosive alkaline substance used in making soap and cleaning plumbing. It is obtained by leaching wood ashes and can be used as a deflocculant for making terra sigillata, though other deflocculants are preferrable. It was probably used by ancient Greek and Roman potters to make their terra sigillata ware

Melon pot – a traditional form among Pueblo potters, featuring vertical lobes or ribs, resembling some kinds of melon and squash

Miracle-Gro – fertilizer containing many chemicals such as copper sulphate which can be used for color in saggar- and pit-firing

Oxidation – firing in an oxygen-rich atmosphere

Paperclay – a clay body with paper pulp added. Commercially manufactured paperclay can be purchased, but many potters add paper pulp to clay slip to create their own. This kind of clay is easy to work with, resists cracking in the drying process, and has great resistance to thermal shock

Polychrome – refers to burnished pottery decorated with multiple colors of slip

Pop-out – also called 'lime pops', a fault of some clays when fired to very low temperatures in which particles of calcium carbonate, or lime, in the clay absorb water after the firing, swelling and causing bits of the surface to pop off

Porcelain – a vitreous high fire white clay body

Pueblo – Native American peoples of the Southwestern United States, who live in established 'villages' or 'Pueblos.' The largest and best-known Pueblos are in the state of New Mexico. Pueblo potters have refined the art of making coil-built burnished pottery to a high degree of perfection and their pots fetch prices in the tens of thousands of dollars

Pyrometer – a thermometer designed to measure high temperatures such as in a kiln

Raku – a type of firing involving rapidly heating pots to a temperature of around 980–1100°C (1800–1900°F), then removing them from the kiln and putting them in a reduction barrel full of combustible material to blacken the unglazed portions of the pot

Reduction – a firing atmosphere which is oxygen poor

Reduction barrel – the container into which hot pots are placed along with combustible materials in the raku process

Saggar – a lidded container used to contain and isolate a pot during its firing

Sgraffito – decorating by scratching, usually with a tool, often through a layer of slip

Slip – any mixture of clay and water

Sodium silicate – a deflocculant used for making terra sigillata, also commonly used to make casting slip

Specific gravity – the density of a substance relative to the density of water

Stoneware – 'high fire' pottery, or clay which matures (becomes non-porous) above about cone 6 (about 1222°C/2232°F)

Terra sigillata – a very fine-textured slip which becomes shiny, or burnished, when it is gently rubbed with a soft material

Thermal shock – cracking due to rapid temperature change

Tumble-stacked – a way of loading a kiln in which pots are stacked on top of each other without the use of kiln furniture

Vermiculite – lightweight insulating material consisting of mica which is expanded by heating, commonly used in potting mixes

Vitrify – to fire clay to the point at which it is no longer porous

Index

(**Bold** indicates images)

acrylic spray glaze, 103
African pottery, 12-14, **13**
Americas, development of indigenous pottery, 14-15
Asselberghs, Wally, 28, 95, 97, **97**, **98**
Axner Super Kiln, 52

Barnes, Edgworth, 79, **79**, **80**, **92**
black-firing,
- in aluminum foil, 51
- in an electric kiln, 52
- in a gas kiln, 54, **56**
- in an oil drum, 52-54, **55**
- in a trash can, 51-52, **53**
- in Mata Ortiz, 54-56, **57**
- Pueblo Indian technique, 58, **59**

burnishing,
- high-talc earthenware clay, 20-22
- leather-hard or black-hard clay, 25-27
- on the wheel, 27-28
- stoneware clay, 17
- technological advantages of, 7, 11
- tools for, 17, 20, 28
- with graphite, 22-24

Burton, Jane, **66**, 67

Carcia, Joan, **28**, 78, **78**, 86
Chinese pottery, 12
clay types, 20, 24, 28-30, 32-33
CMC, 45, 98-99
copper carbonate, 63, 64, 68, 71, 73, 74, 76, 78, 79, 84
crocus martis, 44, **100**

Dadisman, Linda, **40**, **41**, 74, **76**, **77**, **102**
deflocculant, 35-36

Egyptian Pottery, **13**

ferric chloride, 75, 79-80, 84
ferric sulphate, 70

Garrett, Ian, **24**, **25**, 54, **61**, 64, **64**, **65**
gold leaf, 104
graphite, 22, **24**
Greek pottery, 12-13, 35, 110
Greenbaum, David, 27, **27**, 28, **50**, 52, **54**

Hanna, Ashraf, **47**, **61**
Hensler, Bill, 57, 58
horsehair pottery
- alternatives to horsehair, 89, 94, **94**
- firing, 89-92
- images of, **88**, **91-94**
- sources of hair, 92

Kappa, Carla, **30**, **103**
Koch, Gabriele, **2**, **30**, 30

Lee, Jan, 42, 44, **86**, 95, **96**, 97

Maldonado, Ricky, **43**, **44**, 45
Martinez, Maria, **10**
Mata Ortiz, 20, 22, 24, 54, 56, 57, 58
McRaine, Dean, 73
Mediterranean pottery, 12-13
Miracle-Gro, 70, 75
Motzkin, Judith, **47**, **83**, **84**

Naked Raku,
- firing, 95-100,
- naked raku work, **9**, **96**, **98**, **99**, **100**

Norori, Blodomir, **29**

paint, 103
pit-firing,
- firing, 63-73,
- pit-fired work, **8**, **25**, **26**, **27**, **34**, **46**, **50**, **54**, **62**, **66**, **68**, **73**
- sources of potassium for color, 69-70
- sources of sodium for color, 69, 71

polychrome pottery firing, 56-58, **58**

Prier, Carol Molly, 26, **26**, 67, **67**, **68**
Pueblo pottery, **10**, 12, 15, 17, 22, **23**, 58, 60, **60**, **111**

Quezada, Juan, 20, 54
Quezada, Lydia, 57

raku overglaze recipe, 97
Riggs, Charles and Linda, **9**, 42, **45**, 74, **75**, **93**, **99**

Rivera, Ron, 24
Roberts resist slip recipe, 97
Ross, Duncan, **49**

saggar firing,
firing, 51, 74-87
in aluminum foil, 79-80, 84
making saggars, 74
materials for, 74, 75, 78, 79, 84
saggar-fired work, **28**, **41**, **45**, **47**, **61**, **75**, **77**, **78**, **81**, **83**, **85**, **86**, **87**
Salmon, Antonia, **72**, **104**
sgraffito, **9**, **29**, 95
Smith, Bob, **81**, **82**, **87**
specific gravity, measuring, 44

Talavera, Rito, **57**
terra sigillata,
applying, 37-40
making, 35-37
origin of, 13
painting intricate patterns with, 46
recipes, 42-44
spraying, 40
trouble-shooting, 45
tile sealer, 101
Thomason, Travis, **89**, **90**, 90, **91**, **92**

varnish, 102
vermiculite, 52
Villalba, Sabino, **58**
von Dassow, Sumi, **8**, **16**, **48**, **62**, **73**, **85**, **88**, **95**, **105**

Wally's slip recipe, 95, 97
waxes, 101
Wisner, Michael, 20, 22, **22**, **23**, 24, 52-56, **55**, **56**

Youngblood, Nancy, **22**, 23, 58, **59**, **60**, **112**

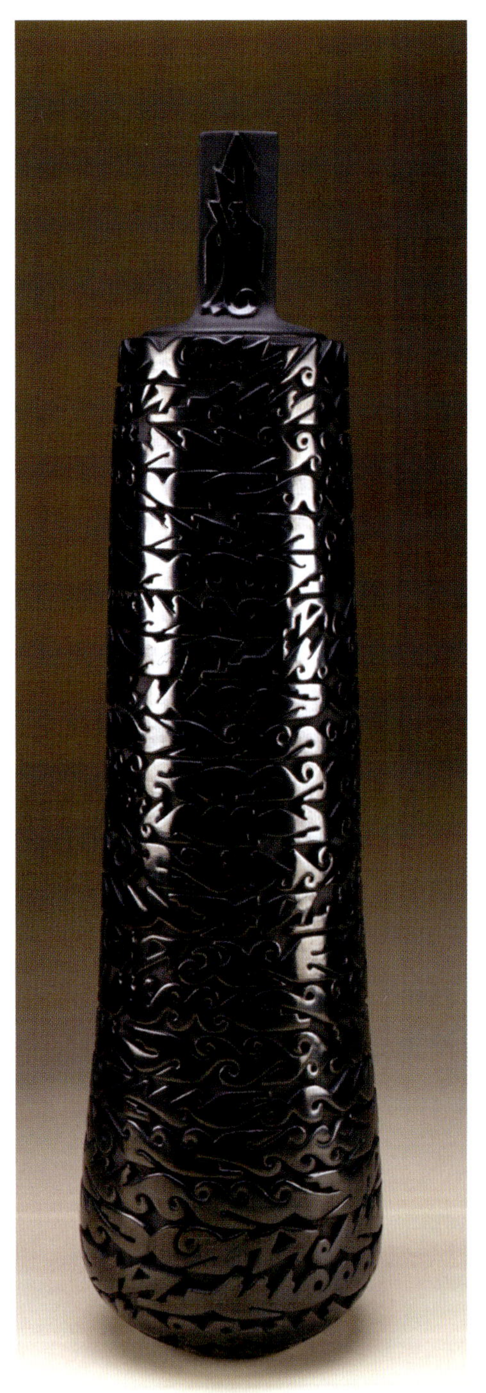

Twenty-four Serpent Vase with Lid by Nancy Youngblood of Santa Clara Pueblo. Coil-built from clay collected on Pueblo land, carved, coated with red slip, burnished with a stone, and smoke-fired. Ht: 55.5 cm (22½ in.), dia: 14 cm (5½ in.), 2002. *Photograph by Hawthorne Studio, Santa Fe NM.*